TGFBI-related Corneal Dystrophies

Eung Kweon Kim • R. Doyle Stulting
Editors

TGFBI-related Corneal Dystrophies

Clinical Findings, Cell Biology, and Genetics

Springer

Editors
Eung Kweon Kim
Department of Ophthalmology
Yonsei University College of Medicine
Seoul, Korea (Republic of)

The Corneal Dystrophy Research Institute
Yonsei University College of Medicine
Seoul, Korea (Republic of)

Saevit Eye Hospital
Goyang, Korea (Republic of)

R. Doyle Stulting
Department of Ophthalmology
Emory University
Atlanta, GA, USA

Woolfson Eye Institute
Atlanta, GA, USA

ISBN 978-981-96-0130-1 ISBN 978-981-96-0131-8 (eBook)
https://doi.org/10.1007/978-981-96-0131-8

This Springer imprint is published by the registered company Springer Nature Singapore Pte Ltd.
The registered company address is: 152 Beach Road, #21-01/04 Gateway East, Singapore 189721, Singapore

Preface

What Is Transforming Growth Factor Beta-Induced Protein?

Transforming growth factor beta-induced protein (TGFBIp) is composed of 683 amino acids in five domains, including the cysteine-rich domain of periostin and TGFBIp (CROPT) on the N-terminal side and four FAS1 domains (Han et al. 2016). Its structure can be divided into a left portion consisting of CROPT, FAS1–1 and FAS-1-2 domains, and a right portion comprising FAS1–3 and FAS1–4 domains (Garcia-Castellanos et al. 2017). Although the quaternary structure of TGFBIp has not been fully elucidated, dimers, trimers, and higher order multimers have been observed. These oligomers exist because TGFBIp has an inherent tendency to aggregate (Basaiawmoit et al. 2011; Kim et al. 2002).

The DNA sequence of transforming growth factor beta-induced (*TGFBI*) gene derived from a human lung adenocarcinoma cell line was first reported in 1992 (Skonier et al. 1992). TGFBIp is found in a variety of organs and tissues, including heart, breast, intestine, liver, skin, kidney, bones, and plasma. In 1994, Klintworth et al. reported that TGFBIp is expressed in all layers of the human cornea, especially the epithelium (Klintworth et al. 1994).

TGFBIp plays numerous physiological and pathological roles in our body. It promotes cell adhesion by binding to integrins and extracellular matrix molecules. It also modulates cellular functions by interacting with proteins in the extracellular matrix. In addition, TGFBIp has been reported to inhibit angiogenic properties of endothelial cells by direct interaction with $\alpha v\beta 3$ integrin (Son et al. 2013). *TGFBI* is also involved in adipocyte differentiation, and the genetic polymorphism of *TGFBI* is associated with the level of insulin and the body mass index (Park et al. 2005). A recent study showed that *TGFBI* reduces adipose angiogenesis by inhibiting the activity of Src and ERK signaling pathways (Lee et al. 2021).

Since the *TGFBI* gene was first discovered in lung adenocarcinoma, research has been directed toward the possible role of *TGFBI* in cancer cells. It has been reported that *TGFBI* is down-regulated in metastatic lung cancer and ovarian cancer through hypermethylation of promoter (Kang et al. 2010; Shah et al. 2008), and overexpression of *TGFBI* in lung cancer reportedly facilitates its response to chemotherapy (Irigoyen et al. 2010). *TGFBI* is up-regulated in patients with colon cancer and pancreatic cancer, and it provides a survival advantage to patients with gastrointestinal tract cancer (Han et al. 2015).

Despite the various systemic physiological functions of *TGFBI*, disease-causing mutations in TGFBIp appear to affect only corneal tissue (El Kochairi et al. 2006). This is believed to be due to unique structural properties of the cornea, such as its avascularity (Han et al. 2016).

The Role of TGFBIp in the Cornea

Mutations in *TGFBI* lead to the accumulation of abnormal materials in the cornea. These conditions, called *TGFBI*-related corneal dystrophies, were originally classified according to their phenotypes, and subtypes were numbered—e.g., granular corneal dystrophy type I (GCDI), granular corneal dystrophy type II (GCDII), and lattice corneal dystrophy type I (LCDI). More recently, genetic analysis has allowed us to associate phenotypic expression with the genetic basis for the corneal dystrophies. A unified, consistent classification system was recently published by Weiss et al. (2008), and it has been revised as new information has become available (Weiss et al. 2015). Similar phenotypes can be produced by different genetic mutations, and phenotypic expression of the same genetic mutations can vary among individuals, making classification of the corneal dystrophies difficult.

Corneal deposits identified by antibodies to TGFBIp are composed of many different kinds of protein debris, which are still under investigation (Han et al. 2016). Phenotypes can also change with time. This temporal variability makes phenotypic classification of corneal dystrophies even more challenging.

Some experts in the field continue to disagree not only with respect to the classification of *TGFBI* dystrophies, but also about the criteria that should be met before a condition is accepted as a distinct corneal dystrophy. For example, some experts believe that there are not sufficient data to prove that LCD variants such as LCD IIIA and LCD IV exist (Walter Lisch, personal communication, 2024), while others, including the author of Chap. 5, believe existing data are sufficient to prove their existence.

Until now, the R124C mutation of the *TGFBI* gene has universally been recognized as the cause of classic LCD (LCD1). Patel et al. (2010), however, reported three cases of granular and lattice deposits in the cornea of an individual with the R124C mutation, demonstrating that a single *TGFBI* mutation such as R124C can lead to multiple phenotypes, some of which may be generally associated with other mutations such as R124H.

Han et al. (2012) reported GCD2 heterozygotes with extreme variation in phenotypes among individual patients. They also reported that TGFBIp expression by different donor-derived cultured corneal fibroblasts was different. Phenotypes tended to be consistent within families, suggesting that the variable expressivity might be regulated by other genetic factors that could influence the quantity of TGFBIp produced or the function of the synthetic pathway.

Choi et al. (2010) detected twofold differences in the expression of 555 genes between wild-type and homozygous GCD2 primary cultured corneal fibroblasts demonstrating that the R124H mutation can be associated with the expression of numerous genes.

Kim et al. (2011) reported that mitochondrial features and function were altered in mutated GCD2 keratocytes, particularly in older cells. This suggests that alteration of mitochondrial function might also affect the pathogenesis of GCD2.

We now realize that there may not be a one-to-one correspondence between phenotype and genotype, with outside factors modifying the expression of *TGFBI* mutations. In this book, we will use the recent nomenclature and classification according to Weiss et al. (2015) with the knowledge that our current beliefs about *TGFBI* dystrophies may become obsolete in the future as newer data become available.

Seoul, Korea (Republic of) — Eung Kweon Kim
Atlanta, GA, USA — R. Doyle Stulting

References

Basaiawmoit RV, et al. SAXS models of TGFBIp reveal a trimeric structure and show that the overall shape is not affected by the Arg124His mutation. J Mol Biol. 2011;408(3):503–13.

Choi SI, et al. Involvement of TGF-{beta} receptor—and integrin-mediated signaling pathways in the pathogenesis of granular corneal dystrophy II. Invest Ophthalmol Vis Sci. 2010;51(4):1832–47.

El Kochairi I, et al. Systemic investigation of keratoepithelin deposits in TGFBI/BIGH3-related corneal dystrophy. Mol Vis. 2006;12:461–6.

Garcia-Castellanos R, et al. Structural and functional implications of human transforming growth factor beta-induced protein, TGFBIp, in corneal dystrophies. Structure. 2017;25(11):1740–1750.e2.

Han KE, et al. Extremely varied phenotypes in granular corneal dystrophy type 2 heterozygotes. Mol Vis. 2012;18:1755–62.

Han B, et al. The role of TGFBI (βig-H3) in gastrointestinal tract tumorigenesis. Mol Cancer. 2015;14:64.

Han KE, et al. Pathogenesis and treatments of TGFBI corneal dystrophies. Prog Retin Eye Res. 2016b;50:67–88.

Irigoyen M, et al. TGFBI expression is associated with a better response to chemotherapy in NSCLC. Mol Cancer. 2010;9:130.

Kang S, Dong SM, Park NH. Frequent promoter hypermethylation of TGFBI in epithelial ovarian cancer. Gynecol Oncol. 2010;118(1):58–63.

Kim JE, et al. Molecular properties of wild-type and mutant beta IG-H3 proteins. Investig Ophthalmol Vis Sci. 2002;43(3):656–61.

Kim TI, et al. Altered mitochondrial function in type 2 granular corneal dystrophy. Am J Pathol. 2011;179(2):684–92.

Klintworth GK, Enghild JJ, Valnickova Z. Discovery of a novel protein (big-h3) in normal human cornea. Invest Ophthalmol Vis Sci. 1994:35, 1938.

Lee SG, et al. Endothelial angiogenic activity and adipose angiogenesis is controlled by extracellular matrix protein TGFBI. Sci Rep. 2021;11(1):9644.

Park KS, et al. Genetic polymorphisms in the transforming growth factor beta-induced gene associated with BMI. Hum Mutat. 2005;25(3):322.

Patel DA, et al. Granular and lattice deposits in corneal dystrophy caused by R124C mutation of TGFBIp. Cornea. 2010;29(11):1215–22.

Shah JN, et al. Methylation screening of the TGFBI promoter in human lung and prostate cancer by methylation-specific PCR. BMC Cancer. 2008;8:284.

Skonier J, et al. cDNA cloning and sequence analysis of beta ig-h3, a novel gene induced in a human adenocarcinoma cell line after treatment with transforming growth factor-beta. DNA Cell Biol. 1992;11(7):511–22.

Son HN, et al. Multiple FAS1 domains and the RGD motif of TGFBI act cooperatively to bind αvβ3 integrin, leading to anti-angiogenic and anti-tumor effects. Biochim Biophys Acta. 2013;1833(10):2378–88.

Weiss JS, et al. The IC3D classification of the corneal dystrophies. Cornea. 2008;27 Suppl 2(Suppl 2):S1–83.

Weiss JS, et al. IC3D classification of corneal dystrophies—edition 2. Cornea. 2015;34(2):117–59.

Contents

1 History of Corneal Dystrophy Focusing on *TGFBI*-Related Corneal Dystrophy

Tae-im Kim, Eung Kweon Kim, and R. Doyle Stulting

The term "corneal dystrophy" (CD) refers to a group of rare hereditary non-inflammatory disorders characterized by abnormal deposition of substances in the cornea (Weiss et al. 2015a), first introduced in 1890 by Arthur Groenouw, who used the term, "Noduli Corneae" (Groenouw 1890). Because the slit lamp was not invented until 20 years later and microscopic examination of the cornea was not possible at the time, early descriptions of corneal dystrophies were not always accurate. Arthur Groenouw, Ernst Fuchs (1910), Wilhelm Uhthoff (1915), and Yoshiharu Yoshida (1924) provided an early definition of corneal dystrophies as bilateral, symmetric, progressive, corneal opacities that are unrelated to environmental factors and systemic abnormalities.

Bücklers (1938) provided the first classification of the corneal dystrophies when he described the differences between granular, lattice, and macular corneal dystrophies. Before the genetic basis of corneal dystrophies was discovered, they were classified on the basis of their location and appearance (American Academy of Ophthalmology 2007). Anatomically, they were classified as epithelial and subepithelial, Bowman's layer, stromal, Descemet membrane, and endothelial dystrophies (Duke-Elder and Leigh 1965; Waring et al. 1978; Klein and Franceschetti 1964; Klintworth 2003a). Deposits may be localized or diffuse, central, or peripheral. They may have characteristic appearances (crumb-like, needle-like, etc.). The stroma between deposits may be clear, cloudy, or opaque. Dystrophies, which are inherited or result from a de novo genetic mutation, are distinguished from degenerations, which are not inherited and result from aging changes (Lin et al. 2016). When the first report of the genetic mutation was published, our classification of corneal dystrophies changed from one based on physical appearance to one based on the responsible genetic mutation.

T.-i. Kim (✉)
Department of Ophthalmology, Yonsei University College of Medicine, Seoul, Korea (Republic of)

The Corneal Dystrophy Research Institute, Yonsei University College of Medicine, Seoul, Korea (Republic of)

Institute of Vision Research, Yonsei University College of Medicine, Seoul, Korea (Republic of)
e-mail: tikim@yuhs.ac

E. K. Kim
Department of Ophthalmology, Yonsei University College of Medicine, Seoul, Korea (Republic of)

The Corneal Dystrophy Research Institute, Yonsei University College of Medicine, Seoul, Korea (Republic of)

Saevit Eye Hospital, Goyang, Korea (Republic of)
e-mail: eungkkim@yuhs.ac

R. D. Stulting
Department of Ophthalmology, Emory University, Atlanta, GA, USA

Woolfson Eye Institute, Atlanta, GA, USA
e-mail: dstulting@icloud.com

E. K. Kim, R. D. Stulting (eds.), *TGFBI-related Corneal Dystrophies*,
https://doi.org/10.1007/978-981-96-0131-8_1

The International Committee for Classification of Corneal Dystrophies (IC3D) was created in 2005 in order to revise the corneal dystrophy nomenclature and create a current and accurate corneal dystrophy classification system (Weiss et al. 2015a). The IC3D categorized CD into four different categories, based on our genetic understanding of each disease variant: Category 1 identifies a CD with the mutated gene(s) identified; Category 2 identifies the locus of the mutation, category 3 is a CD without a gene or locus identified, and category 4 is reserved for new or suspicious CD (Weiss et al. 2015a; Vedana et al. 2016). Many dystrophies are well understood genetically and are, therefore, in category 1. Lisch epithelial CD and X-linked endothelial corneal dystrophy (XECD) are in category 2. Others, such as epithelial recurrent erosion dystrophies (ERED) and posterior amorphous CD (category 3) and subepithelial mucinous CD (category 4), are not yet well understood genetically (Klintworth 2009). The goal was to establish a new nomenclature that would reflect current clinical, pathologic, and genetic knowledge, be easily adaptable to advances in understanding from the continued discovery of new genes and mutations, and be linked to the old nomenclature for ease of use.

Transforming growth factor beta-induced (*TGFBI*) corneal dystrophies were included in the section on stromal dystrophies in the IC3D classification initially published in 2008 (Weiss et al. 2015a). In addition to clinical photographs included in the first edition, pathology images were included in a second edition in 2015 (Weiss et al. 2015b). Congenital hereditary endothelial dystrophy (CHED) was reclassified and corneal dystrophy nomenclature became more guided by genetics. All corneal dystrophies known to map to the *TGFBI* gene were included in a single independent section designated as epithelial–stromal *TGFBI* dystrophies in IC3D 2nd ed. (2015) apart from the section of stromal dystrophies in which *TGFBI*-related corneal dystrophies had been included in IC3D ed. (2008). The third edition of IC3D is now being revised based on peer-reviewed publications from 2014 to 2023. For the LCD variant with non-R124C *TGFBI* mutations, a new nomenclature for specific LCD pathogenic variants would be proposed by appending the mutation using 1-letter amino acid abbreviations to LCD.

The transforming growth factor beta-induced (*TGFBI*) gene, found on chromosome 5 and the function of the protein it produces is to link to collagens and provide support in the corneal extracellular matrix (Ween et al. 2012). The structure and function of this protein is altered in *TGFBI*-related corneal dystrophies such as type LCD type 1, GCD types 1 and 2, Reis-Bücklers CD, and Thiel-Behnke CD with an autosomal dominant pattern (Lin et al. 2016; Klintworth 2009; Ween et al. 2012; Aldave and Sonmez 2007).

The description and classification of CD began before the era of the slit lamp, and tremendous progress has been made in understanding the pathogenesis and genetic abnormalities. Our advanced, modern understanding of CD may someday provide a better management of the CD for families for which we do not have an effective treatment strategy today.

References

Aldave AJ, Sonmez B. Elucidating the molecular genetic basis of the corneal dystrophies: are we there yet? Arch Ophthalmol. 2007;125(2):177–86.

American Academy of Ophthalmology. External diseases and cornea. In: Sutphin JE, editor. Basic and clinical sciences course 2007–2008. San Francisco, CA: American Academy of Ophthalmology; 2007. p. 305–29.

Bücklers M. Die erblichen hornhaut-dystrophie. Klin Monatsbl Augenheilkd. 1938;3:1–135.

Duke-Elder S, Leigh AG. Corneal dystrophies. In: Duke-Elder S, editor. System of ophthalmology. Part 2, vol. 8. London: Kimpton; 1965. p. 864–7.

Fuchs E. Dystrophia epithelialis corneae. Albrecht von Graefe's Arch Clin Exp Ophthalmol. 1910;76: 478–508.

Groenouw A. Knötchenförmige hornhauttrübungen "Noduli Corneae". Arch Augenheilkd. 1890;21: 281–9.

Klein D, Franceschetti A. Heredo-familiäre Hornhautdystrophie. In: Becker PD, editor. Humangenetik, vol. 4. Stuttgart: Georg Thieme; 1964. p. 80–1.

Klintworth GK. The molecular genetics of the corneal dystrophies—current status. Front Biosci. 2003a;8:687–713.

Klintworth GK. Corneal dystrophies. Orphanet J Rare Dis. 2009;4:7.

Lin ZN, Chen J, Cui HP. Characteristics of corneal dystrophies: a review from clinical, histological and genetic perspectives. Int J Ophthalmol. 2016;9(6):904–13.

Uhthoff W. Ein Fall von doppelseitiger zentraler, punktförmiger, supepithelialer knötchenförmiger Keratitis. Groenouw mit anatomischem Befunde. Klin Monatsbl Augenheilkd. 1915;54:377–83.

Vedana G, Villarreal G, Jun AS. Fuchs endothelial corneal dystrophy: current perspectives. Clin Ophthalmol. 2016;10:321–30.

Waring GO, Rodrigues MM, Laibson PR. Corneal dystrophies. I. Dystrophies of the epithelium, Bowman's layer and stroma. Surv Ophthalmol. 1978;23:71–122.

Ween MP, Oehler MK, Ricciardelli C. Transforming growth Factor-Beta-Induced Protein (TGFBI)/(βig-H3): a matrix protein with dual functions in ovarian cancer. Int J Mol Sci. 2012;13(8):10461–77.

Weiss JS, Møller HU, Aldave AJ, Seitz B, Bredrup C, Kivelä T, Munier FL, Rapuano CJ, Nischal KK, Kim EK, Sutphin J, Busin M, Labbé A, Kenyon KR, Kinoshita S, Lisch W. IC3D classification of corneal dystrophies—edition 2. Cornea. 2015a;34(2):117–59.

Weiss JS, Møller H, Aldave A, et al. The IC3D classification of the corneal dystrophies-edition 2. Cornea. 2015b;34:117–59.

Yoshida Y. Über eine neue art der dystrophia corneae mit histologischem Befunde. Albrecht von Graefe's Arch Clin Exp Ophthalmol. 1924;114:91–100.

TGFBI Corneal Dystrophies: Reis—Bücklers Corneal Dystrophy (RBCD)

2

Kyung Eun Han, Woo Chan Park, and R. Doyle Stulting

Reis—Bücklers corneal dystrophy (RBCD) is caused by an autosomal dominantly inherited mutation in the transforming growth factor beta-induced (*TGFBI*) gene, R124L (Arg124Leu) in chromosome 5q31.

In 2008, The International Classification of Corneal Dystrophies (IC3D) classified RBCD and Thiel Behnke corneal dystrophy (TBCD) as Bowman layer corneal dystrophies in which deposits accumulate primarily in Bowman layer and the subepithelial stroma (Weiss et al. 2008). Because the corneal deposits in RBCD and TBCD are also found in the mid—to posterior stroma (Qiu et al. 2016; Weiss et al. 2015), these corneal dystrophies were reclassified in IC3D as epithelial-stromal *TGFBI* dystrophies.

Several alternative names have been used to describe these conditions, such as corneal dystrophy of Bowman layer type I (CDB I), geographic corneal dystrophy (Weidle), atypical granular corneal dystrophy, granular corneal dystrophy type 3, anterior limiting membrane dystrophy type 1, and superficial granular corneal dystrophy. These alternative names are no longer used.

In 1917, Reis first reported this corneal dystrophy in two female patients in a family (Reis 1917), and in 1949, Bücklers reported similar cases in the same family (Bücklers 1949). TBCD was not separated from RBCD for 28 years, until the report of Thiel and Behnke was published in 1967. Clinically, the differential diagnosis of RBCD and TBCD may be difficult—especially during the first two decades of life. In later years, geographic opacities are seen in RBCD, and honeycomb opacities are seen in TBCD, making the correct diagnosis more obvious. The causative mutation, confocal, and electron microscopic findings are different for these two dystrophies (Kobayashi and Sugiyama 2007; Qiu et al. 2016).

Clinical Manifestations

Bilateral white-to-gray subepithelial corneal deposits appear primarily in Bowman layer in children at 4 to 5 years of age (Bücklers 1949; Reis 1917). Patients usually complain of photophobia, tearing, and intermittent ocular irritation

K. E. Han (✉)
Department of Ophthalmology, Ewha Womans University Mokdong Hospital, Ewha Womans University College of Medicine, Seoul, Korea (Republic of)
e-mail: hanke@ewha.ac.kr

W. C. Park
Department of Ophthalmology, Dong-A University Hospital, Dong-A University College of Medicine, Busan, Korea (Republic of)
e-mail: wcpark@dau.ac.kr

R. D. Stulting
Department of Ophthalmology, Emory University, Atlanta, GA, USA

Woolfson Eye Institute, Atlanta, GA, USA
e-mail: dstulting@icloud.com

E. K. Kim, R. D. Stulting (eds.), *TGFBI-related Corneal Dystrophies*,
https://doi.org/10.1007/978-981-96-0131-8_2

caused by recurrent corneal erosions. The corneal deposits enlarge diffusely but remain separated from the limbus by 1–2 mm until finally reaching the limbus and deeper stroma in older patients.

Anterior Segment Optical Coherence Tomography (AS-OCT) and Confocal Microscope Imaging

AS-OCT reveals reflective homogeneous and band-like deposits (Liang et al. 2014; Qiu et al. 2016) or a sawtooth-like appearance in Bowman layer and anterior stroma (Fogla and Knyazer 2014; Liang et al. 2014).

In vivo confocal microscopy reveals focal reflective materials in the superficial epithelial layer and the epithelial basal cell (Kobayashi and Sugiyama 2007; Liang et al. 2014; Werner et al. 1999). Highly reflective irregular aggregations replace Bowman layer, and grayish aggregations are seen in the anterior and mid-stroma (Qiu et al. 2016). Subbasal nerve plexuses are not apparent. Descemet membrane and endothelium are not affected.

Histologic Analysis

Light microscopy reveals replacement of Bowman layer by abnormal deposits, which stain pink with Congo red, pink with hematoxylin and eosin, bright red with Masson's trichrome, and bright red with Van de Grift stain (Kuchle et al. 1995; Qiu et al. 2016; Weiss et al. 2015). Corneal deposits align as 3 to 5 bands parallel to corneal collagen fibers in the anterior stroma. Dot-like aggregates are scattered in the mid—to posterior stroma and in stromal keratocytes (Weiss et al. 2015), which also stain with anti-transforming growth factor beta-induced protein (TGFBIp) antibody (Qiu et al. 2016).

Transmission electron microscopy reveals rod-shaped or trapezoidal electron-dense bodies similar to those seen in corneas with granular corneal dystrophy type 1, which are 100–500 nm wide and located primarily in the subepithelial and anterior stroma (Kuchle et al. 1995; Winkelman et al. 1986). These bodies are scattered throughout the mid—to posterior stroma and are also observed in the stromal keratocytes (Liang et al. 2014; Qiu et al. 2016).

Treatment

When visual acuity decreases, phototherapeutic keratectomy (PTK) can be used to remove superficial deposits. PTK has several advantages over keratoplasty. It is less invasive and repeatable in virgin and grafted eyes, and it may postpone keratoplasty for an extended period of time (Eggink et al. 2002; Fogla and Knyazer 2014). Because most deposits accumulate in the subepithelial and Bowman layers, therapeutic ablation depths from 15–80 μm are often adequate (Eggink et al. 2002; Weiss et al. 2015). Staged PTK ablation (Jung et al. 2013), using intraoperative slit lamp examinations to determine the adequacy of the ablation or sequential customized therapeutic keratectomy, which uses intraoperative topography, tomography, and corrected distance visual acuity (Vinciguerra et al. 2018) can be used to assure adequate removal of deposits, while minimizing the depth of ablation. Lamellar keratoplasty, deep lamellar keratoplasty (conventional or femtosecond-assisted), or penetrating keratoplasty may eventually be required to restore vision (Caldwell 1978; Olson and Kaufman 1978; Qiu et al. 2016; Steger et al. 2016). Recurrence has been reported after PTK (8/17 eyes [47%]; Dinh et al. 1999), lamellar keratoplasty (Liang et al. 2014), and penetrating keratoplasty (Dinh et al. 1999; Eggink et al. 2002; Kuchle et al. 1995). Recurrence tends to occur earlier and more extensively after penetrating keratoplasty in patients with RBCD (as early as 1 year after RBCD versus 3–8 years after TBCD) than in patients with TBCD (Winkelman et al. 1986). The use of intraoperative topical mitomycin has been suggested to reduce the incidence and severity of recurrences (Miller et al. 2004); however, the long-term safety and efficacy of this treatment modality has yet to be established (Figs. 2.1, 2.2, 2.3, 2.4, 2.5, 2.6, and 2.7).

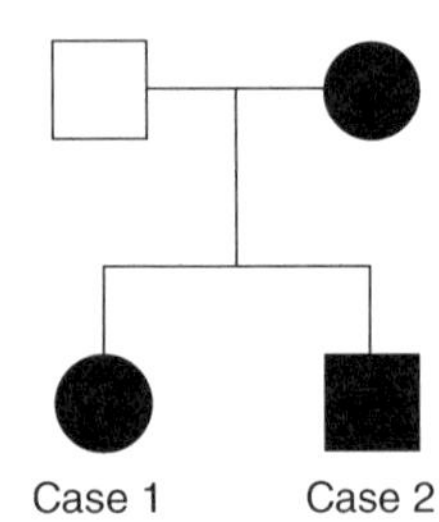

Fig. 2.1 A pedigree of case 1 and 2 affected by Reis–Bücklers corneal dystrophy

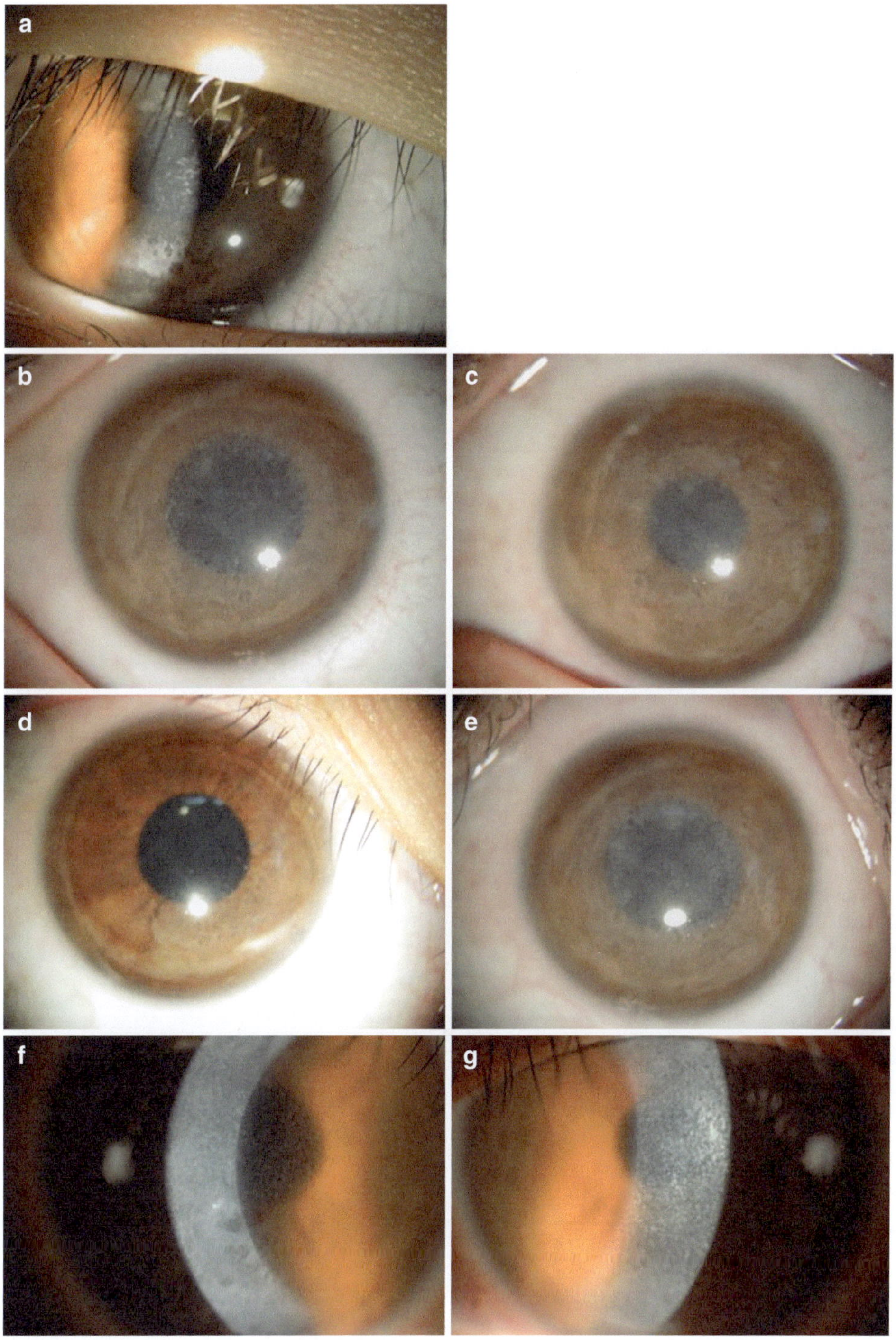

Fig. 2.2 Anterior segment photographs of case 1. Right eye (**b, d, f**), left eye (**a, c, e, g**). An 11-year-old girl complained bilateral corneal opacity and decreased visual acuity in both eyes. She had undergone PTK on the left eye 1 year ago at another hospital (**a**). After 3 years and 6 months, diffuse corneal deposits were observed as a geographic pattern with narrow clear zone from limbus. Her corrected visual acuity was 0.7 in the right eye and 0.16 in the left eye (**b, c**). In the center of the left eye, a recurrence of corneal deposits was observed (**c**). Two months later, she had PTK in the right eye at our clinic (**d**). Corneal opacity in the left eye became denser than the previous visit (**e**). Six years later, a recurrence was observed in both eyes. The corrected visual acuity became 0.05 in the right eye and 0.5 in the left eye (**f, g**)

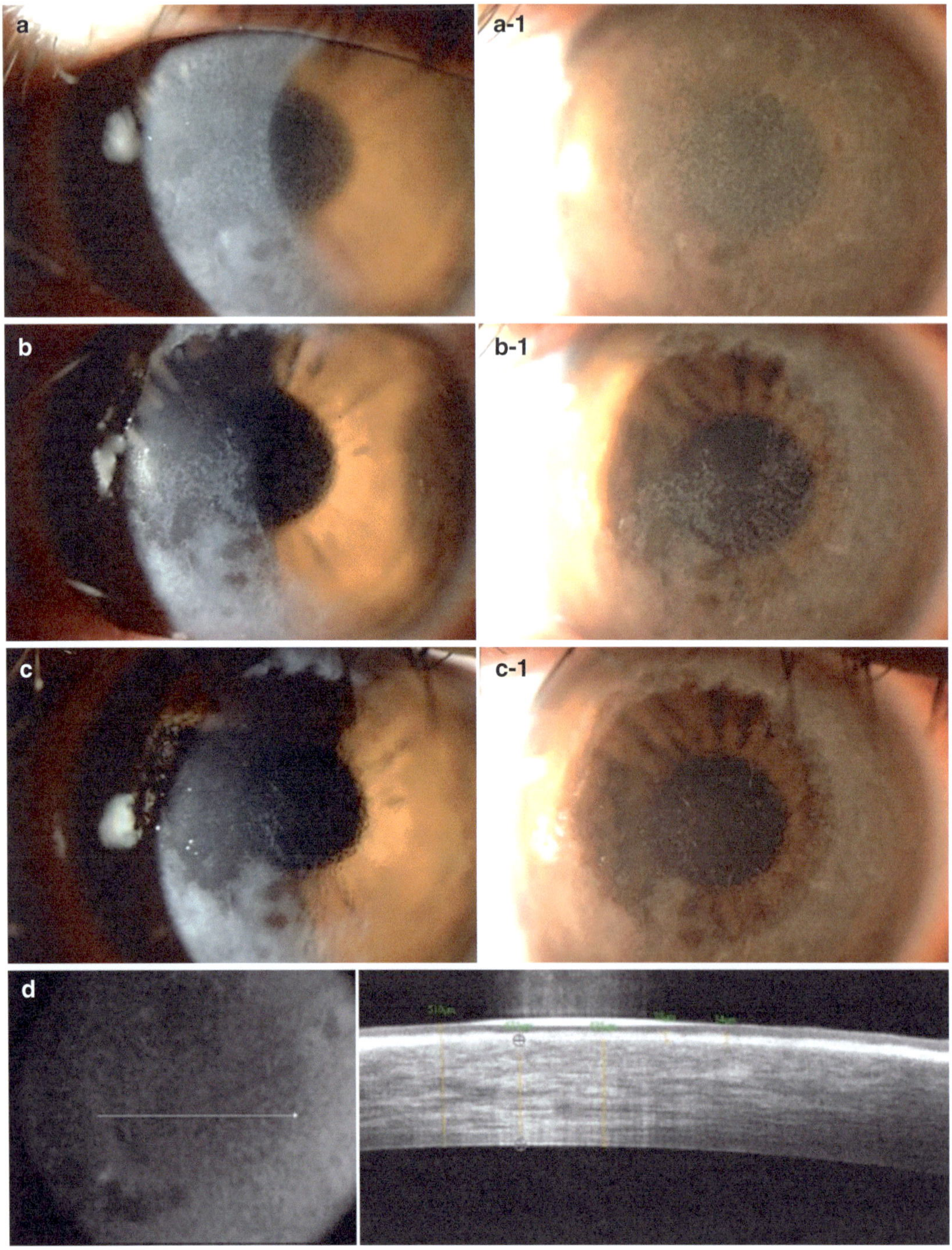

Fig. 2.3 Intraoperative anterior segment photographs and preoperative anterior segment optical coherence tomography (AS-OCT) of the right eye of case 1. She had second PTK procedure (a 43-μm ablation) in the right eye. Preoperative (**a, a-1**). After 34 μm ablation (**b, b-1**). After 43 μm ablation (**c, c-1**). A range of corneal deposit thickness was 40–50 μm in the preoperative AS-OCT (**d**)

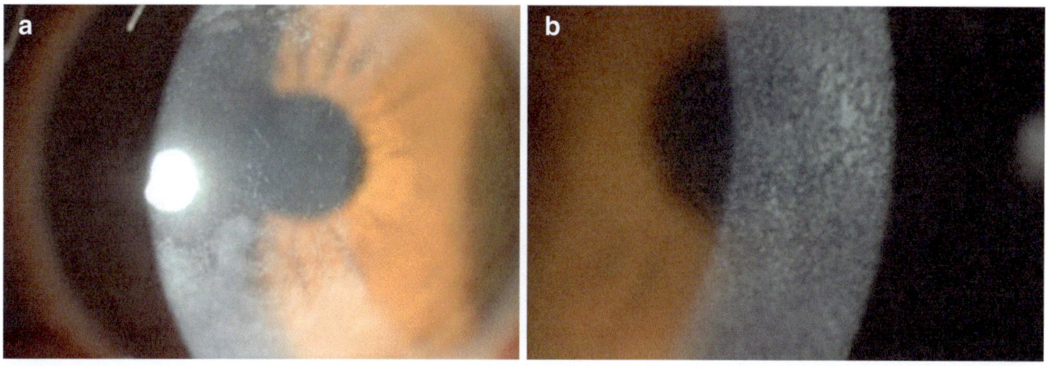

Fig. 2.4 Anterior segment photographs of case 1. Two years later, center of the cornea in the right eye showed minimal recurrence (**a**), but a recurrence was severe in the left eye (**b**)

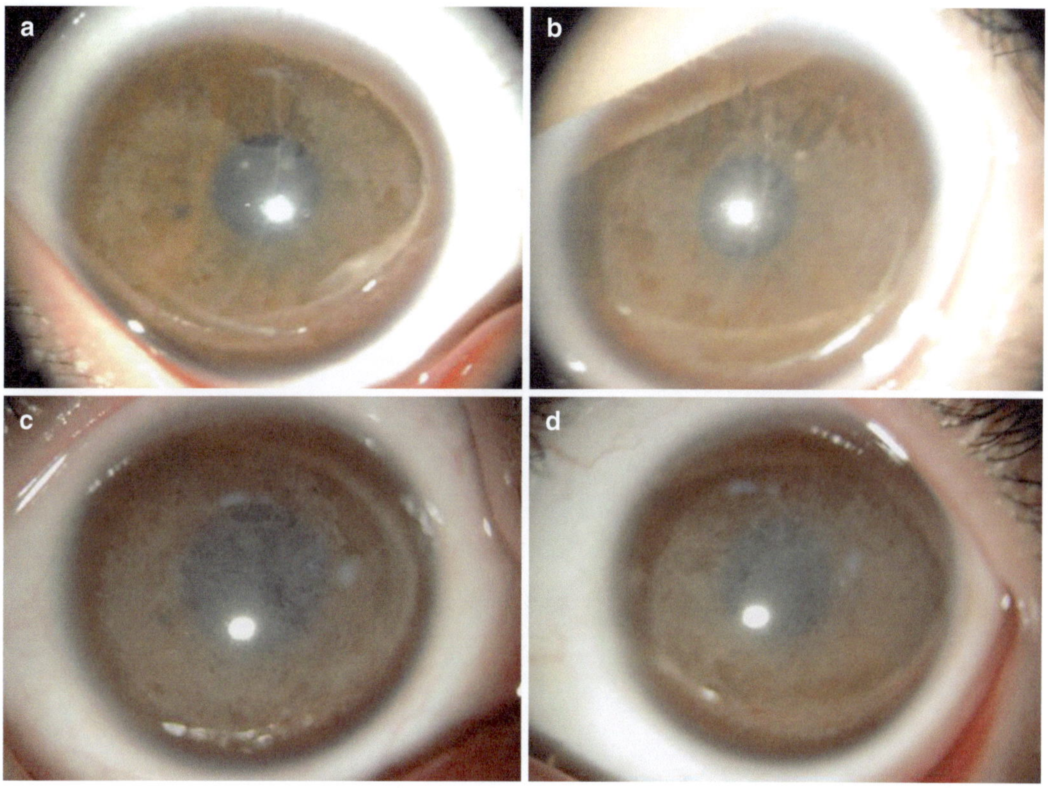

Fig. 2.5 Anterior segment photographs of case 2. Right eye (**a, c, e, g**), left eye (**b, d, f, h**). A 10-year-old boy showed similar corneal opacity to that of his older sister (case 1) (**a, b**). Three years later, bilateral corneal opacity became denser than the previous visit. When he visited our clinic, uncorrected visual acuity was 0.2 in the right eye and 0.3 in the left eye (**c, d**). A PTK (a 30-μm ablation) was performed in his right eye. Three months later, the right eye became clear in the ablated zone. Uncorrected visual acuity in the right eye became 0.6 and uncorrected visual acuity remained 0.3 in the left eye (**e, f**). He revisited to our clinic 8 years later (at age 22). During the follow-up loss period, he had undergone PTK in the left eye at another hospital 4 years ago; however, corneal opacity showed recurrence inside the ablated zone and denser opacity outside the ablated zone. Corrected visual acuity was 0.05 in the right eye and 0.3 in the left eye (**g, h**). In the left eye, some part of corneal deposits spontaneously detached (yellow arrows) (**h**)

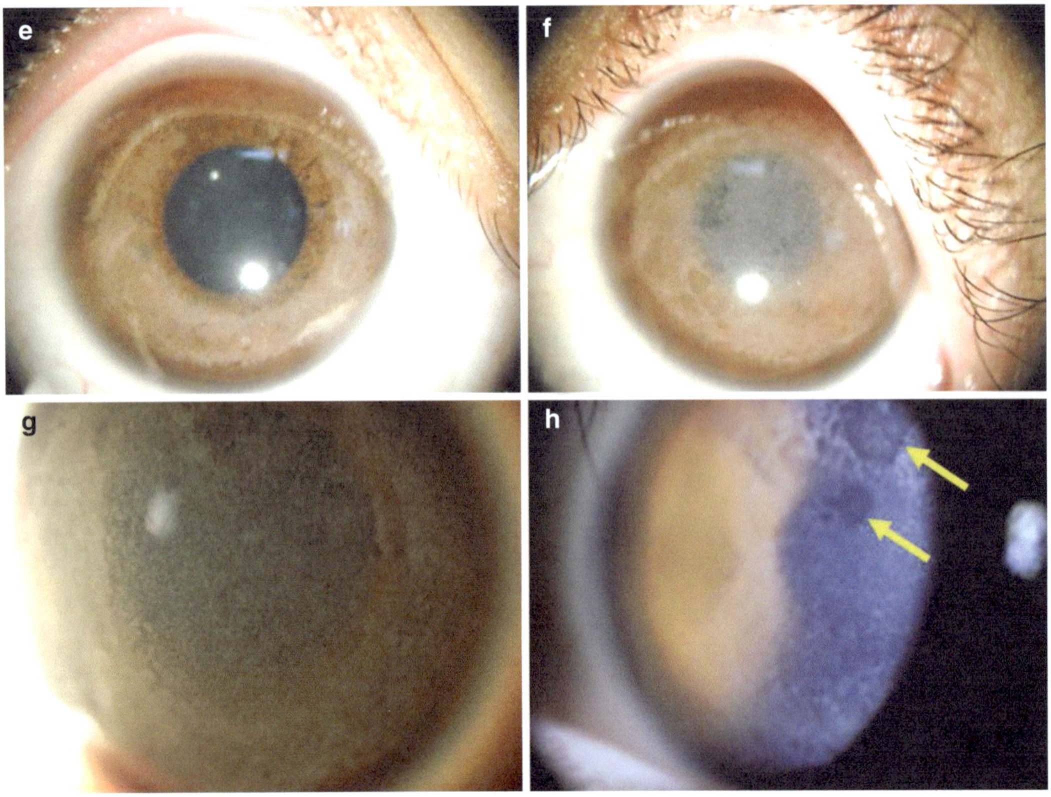

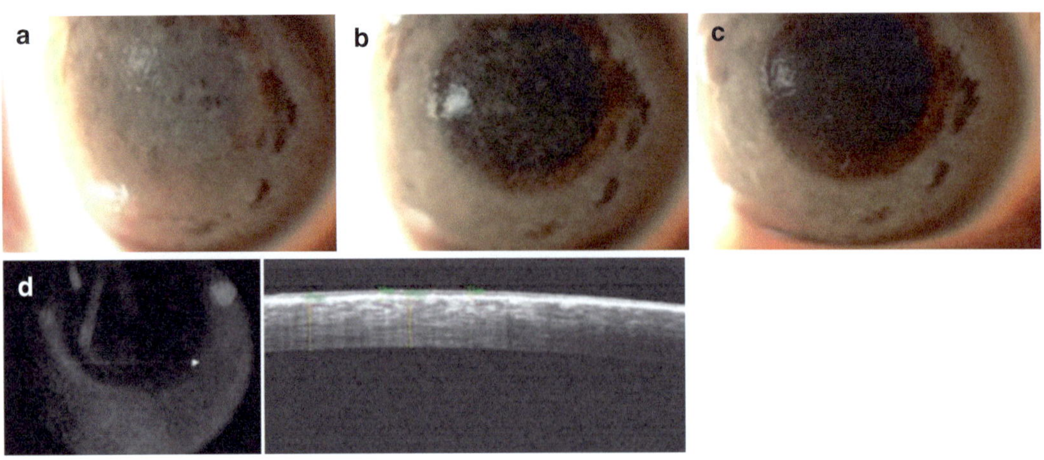

Fig. 2.6 Intraoperative anterior segment photographs and preoperative anterior segment optical coherence tomography (AS-OCT) of the right eye of case 2. He had second PTK (a 40-μm ablation) in the right eye. After epithelial removal, a diffuse corneal opacity was observed (**a**). After 30 μm ablation, corneal deposits in visual axis were remained (**b**). After 40 μm ablation, corneal deposits in visual axis were almost removed (**c**). The mean thickness of corneal deposits was 53 μm in the AS-OCT (**d**)

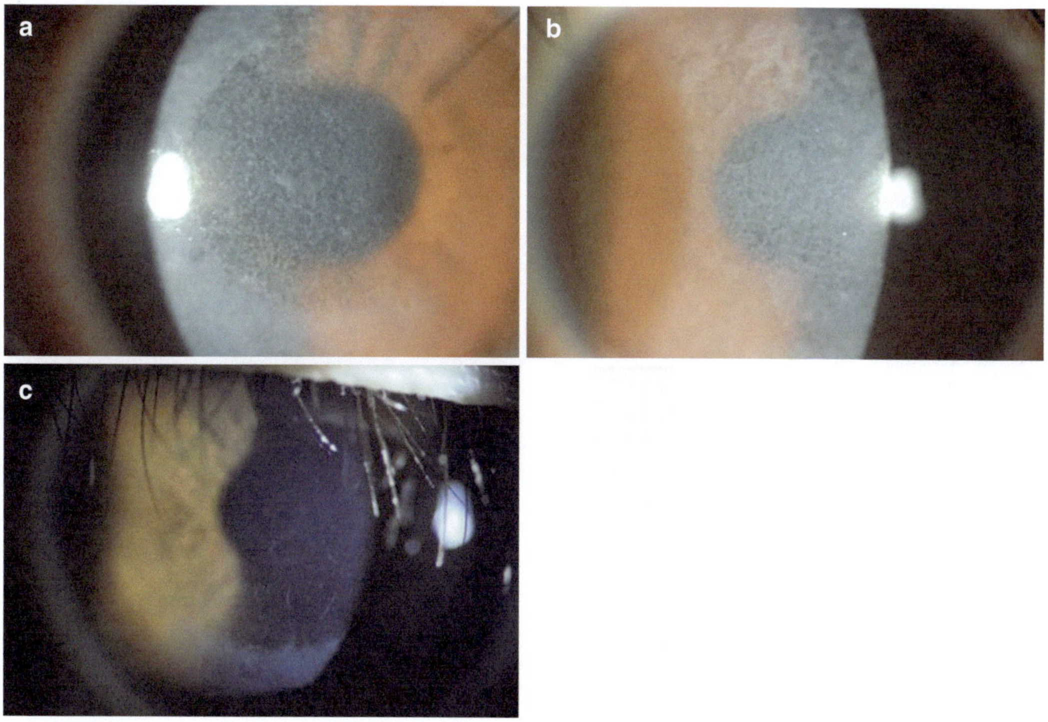

Fig. 2.7 Anterior segment photographs of case 2. Right eye (**a**), left eye (**b, c**). Four years after PTK in the right eye, a recurrence of corneal deposits in the ablation zone was observed in the right eye (**a**) and dense corneal deposits were observed in the left eye (**b**). His uncorrected visual acuity became 0.15 in the right eye and 0.02 in the left eye. He had 2nd PTK in the left eye 1 week later at our clinic. One month later, his left eye became clear in the ablated zone showing several prominent corneal nerve fibers (**c**). Uncorrected visual acuity became 0.4 in the left eye

References

Bücklers M. Uber eine weitere familiare Hornhautdystrophie (Reis). Klin Monatsbl Augenheilkd. 1949;114:386–97.

Caldwell DR. Postoperative recurrence of Reis-Bücklers' corneal dystrophy. Am J Ophthalmol. 1978;85:567–8.

Dinh R, Rapuano CJ, Cohen EJ, Laibson PR. Recurrence of corneal dystrophy after excimer laser phototherapeutic keratectomy. Ophthalmology. 1999;106:1490–7.

Eggink FA, Geerards AJ, Beekhuis WH. Recovery of the visual acuity in a family with Reis-Buckler dystrophy. Cont Lens Anterior Eye. 2002;25:67–72.

Fogla R, Knyazer B. Microkeratome-assisted two-stage technique of superficial anterior lamellar keratoplasty for Reis-Bücklers corneal dystrophy. Cornea. 2014;33:1118–22.

Jung SH, Han KE, Stulting RD, Sgrignoli B, Kim TI, Kim EK. Phototherapeutic keratectomy in diffuse stromal haze in granular corneal dystrophy type 2. Cornea. 2013;32:296–300.

Kobayashi A, Sugiyama K. In vivo laser confocal microscopy findings for Bowman's layer dystrophies (Thiel-Behnke and Reis-Bücklers corneal dystrophies). Ophthalmology. 2007;114:69–75.

Kuchle M, Green WR, Volcker HE, Barraquer J. Reevaluation of corneal dystrophies of Bowman's layer and the anterior stroma (Reis-Bücklers and Thiel-Behnke types): a light and electron microscopic study of eight corneas and a review of the literature. Cornea. 1995;14:333–54.

Liang Q, Pan Z, Sun X, Baudouin C, Labbe A. Reis-Bücklers corneal dystrophy: a reappraisal using in vivo and ex vivo imaging techniques. Ophthalmic Res. 2014;51:187–95.

Miller A, Solomon R, Bloom A, Palmer C, Perry HD, Donnenfeld ED. Prevention of recurrent Reis-Bücklers dystrophy following excimer laser phototherapeutic keratectomy with topical mitomycin C. Cornea. 2004;23:732–5.

Olson RJ, Kaufman HE. Recurrence of Reis-Bücklers' corneal dystrophy in a graft. Am J Ophthalmol. 1978;85:349–51.

Qiu WY, Zheng LB, Pan F, Wang BB, Yao YF. New histopathologic and ultrastructural findings in Reis-Bücklers corneal dystrophy caused by the Arg124Leu mutation of TGFBI gene. BMC Ophthalmol. 2016;16:158.

Reis W. Familiare, fleckige Hornhautentartung. Dtsch Med Wochenschr. 1917;43:575.

Steger B, Romano V, Biddolph S, Willoughby CE, Batterbury M, Kaye SB. Femtosecond laser-assisted lamellar keratectomy for corneal opacities secondary to anterior corneal dystrophies: an interventional case series. Cornea. 2016;35:6–13.

Vinciguerra P, Vinciguerra R, Randleman JB, Torres I, Morenghi E, Camesasca FI. Sequential customized therapeutic keratectomy for Reis-Bücklers' corneal dystrophy: long-term follow-up. J Refract Surg. 2018;34:682–8.

Weiss JS, Moller HU, Lisch W, Kinoshita S, Aldave AJ, Belin MW, Kivela T, Busin M, Munier FL, Seitz B, Sutphin J, Bredrup C, Mannis MJ, Rapuano CJ, Van Rij G, Kim EK, Klintworth GK. The IC3D classification of the corneal dystrophies. Cornea. 2008;27 Suppl 2:S1–83.

Weiss JS, Moller HU, Aldave AJ, Seitz B, Bredrup C, Kivela T, Munier FL, Rapuano CJ, Nischal KK, Kim EK, Sutphin J, Busin M, Labbe A, Kenyon KR, Kinoshita S, Lisch W. IC3D classification of corneal dystrophies—edition 2. Cornea. 2015;34:117–59.

Werner LP, Werner L, Dighiero P, Legeais JM, Renard G. Confocal microscopy in Bowman and stromal corneal dystrophies. Ophthalmology. 1999;106:1697–704.

Winkelman JE, Wittebol-Post D, Delleman JW. Reis-Bücklers corneal dystrophy. Klin Monbl Augenheilkd. 1986;188:143–7.

TGFBI Corneal Dystrophies: Thiel-Behnke Corneal Dystrophy (TBCD)

3

Kyung Eun Han, Yong Woo Ji, and Kyoung Yul Seo

Thiel-Behnke corneal dystrophy (TBCD) is an autosomal dominantly inherited stromal corneal dystrophy associated with an R555Q (Arg555Gln) mutation in the transforming growth factor beta-induced (*TGFBI*) gene on chromosome 5q31. In 1967, Thiel and Behnke first described this corneal dystrophy, in which corneal deposits accumulate primarily in Bowman's layer (Thiel and Behnke 1967). The deposits are different than those of Reis-Bücklers corneal dystrophy (RBCD). TBCD was also called as Grayson-Wilbrandt corneal dystrophy and corneal dystrophy Bowman 2 (CDB2) at one time, but these terms are no longer used.

The 10q-linked TBCD phenotype was first reported in 1997 (Sorour et al. 2005; Yee et al. 1997) as a variant of TBCD. However, the c.3156C > T or c.2816 C > T mutations in the *COL17A1* gene on chromosome 10q23–24 are responsible for epithelial recurrent erosion dystrophy (MIM#122400), and the 10q-liked TBCD is no longer considered as a variant of TBCD (Lisch and Kivela 2013; Lisch and Weiss 2019; Sorour et al. 2005; Sullivan et al. 2003; Weiss et al. 2015) (Figs. 3.1, 3.2, 3.3, 3.4, 3.5, 3.6, and 3.7).

K. E. Han (✉)
Department of Ophthalmology, Ewha Womans University Mokdong Hospital, Ewha Womans University College of Medicine, Seoul, Korea (Republic of)
e-mail: hanke@ewha.ac.kr

Y. W. Ji
Department of Ophthalmology, Yongin Severance Hospital, Yonsei University College of Medicine, Yongin, Korea (Republic of)
e-mail: lusita30@yuhs.ac

K. Y. Seo
Department of Ophthalmology, Institute of Vision Research, Severance Hospital, Yonsei University College of Medicine, Seoul, Korea (Republic of)
e-mail: seoky@yuhs.ac

E. K. Kim, R. D. Stulting (eds.), *TGFBI-related Corneal Dystrophies*,
https://doi.org/10.1007/978-981-96-0131-8_3

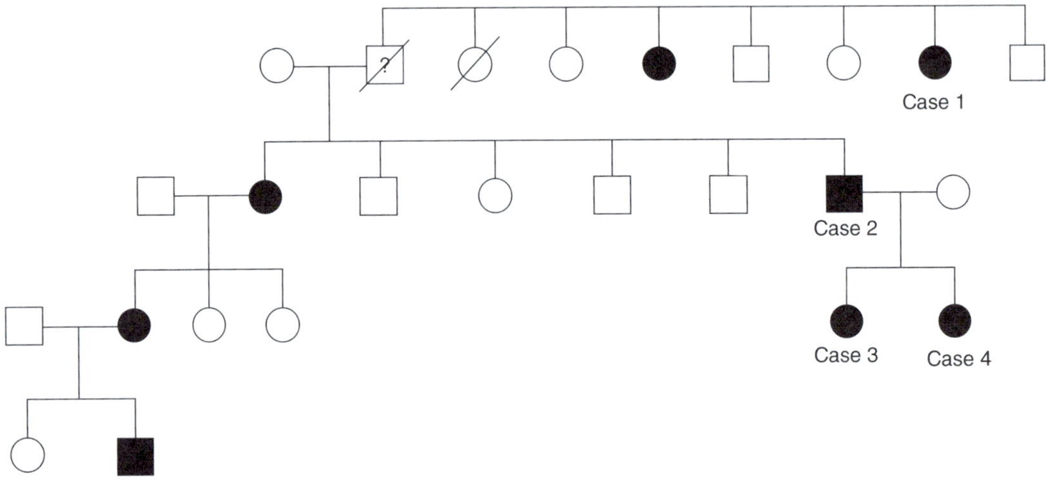

Fig. 3.1 A pedigree of a family with Thiel-Behnke corneal dystrophy (Cases 1–4)

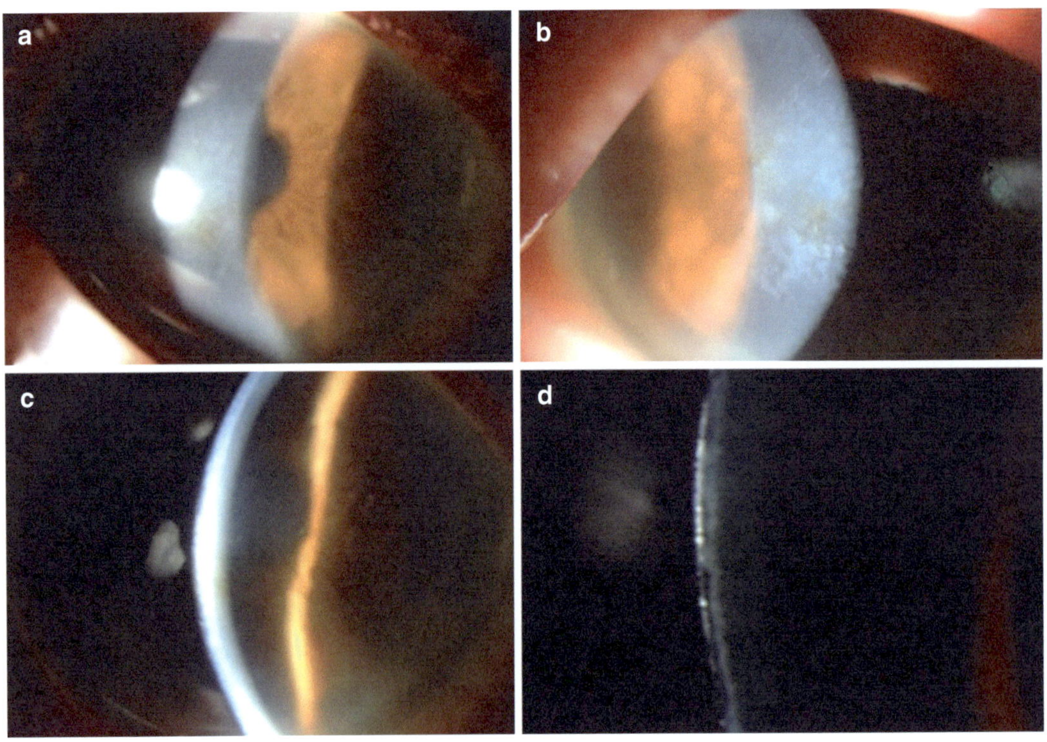

Fig. 3.2 Anterior segment photographs of case 1. Right eye (**a, c, d**), left eye (**b**). A 75-year-old female complained of decreased visual acuity, and corneal opacities were observed in both eyes. Uncorrected visual acuity was 0.1 in the right eye and 0.16 in the left eye. Diffuse central opacities and a brown iron deposit inferior to the central opacity were observed. Thin slit illumination (**c, d**) revealed sawtooth-like corneal deposits in the subepithelial stroma and Bowman layer

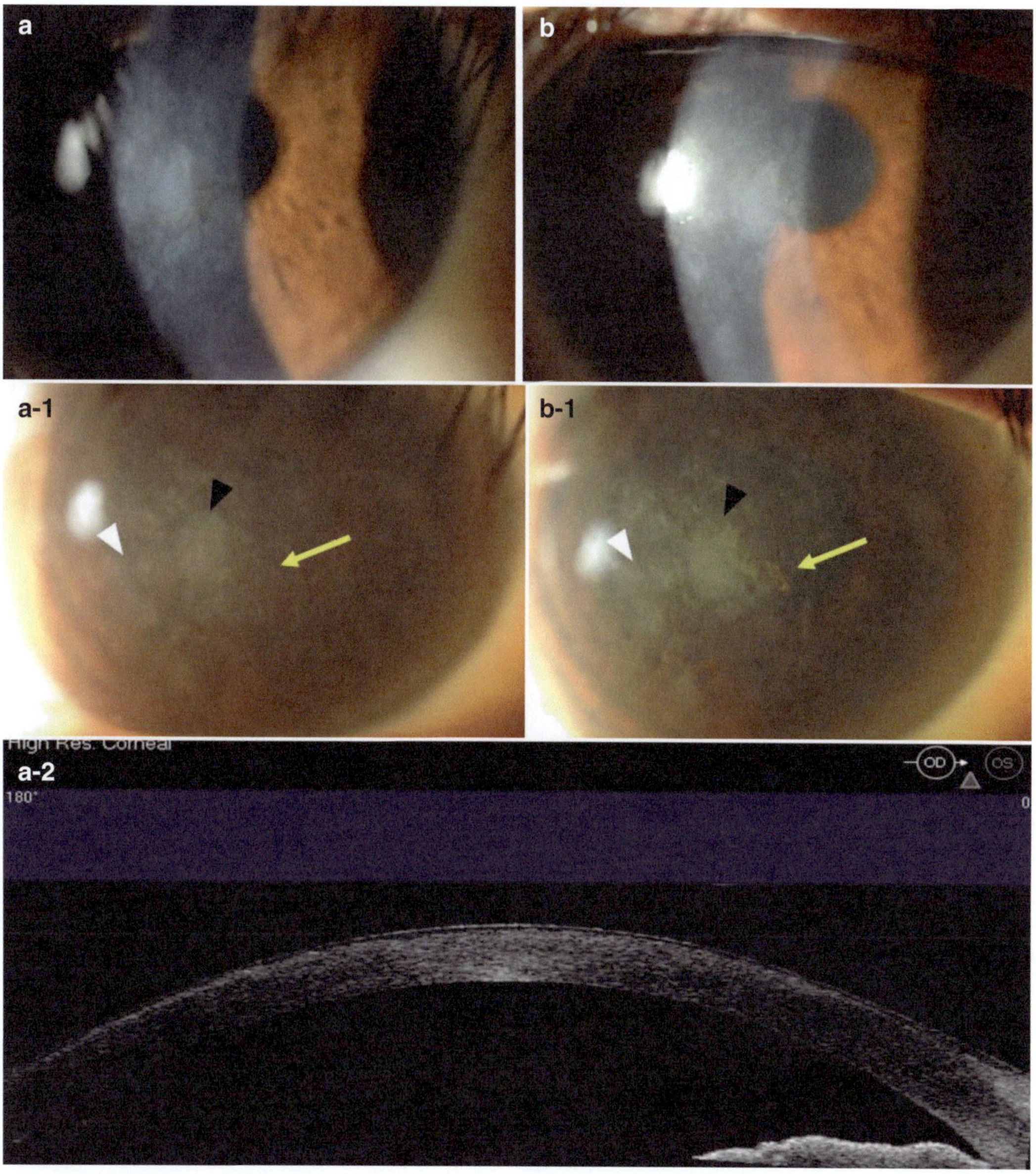

Fig. 3.3 Anterior segment photographs and anterior segment optical coherence tomography (AS-OCT) of the right eye of case 2. At presentation (**a, a-1, a-2**) and 7 years later (**b, b-1**). A 42-year-old male presented with a previous diagnosis of TBCD. Best-corrected visual acuity was 0.9 in the right eye. At presentation, central opacities and an accompanying iron line were observed (**a, a-1**). AS-OCT revealed irregular sawtooth-like reflective opacities in the subepithelial stroma and Bowman layer (**a-2**). Seven years later, one central opacity became denser (white arrowhead), while other corneal opacities became less dense (black arrowhead), and a yellow-brownish scar appeared (yellow arrow) (B and B-1). The white arrowhead, black arrowhead, and yellow arrow indicate the same location in (**a-1**) and (**b-1**), respectively

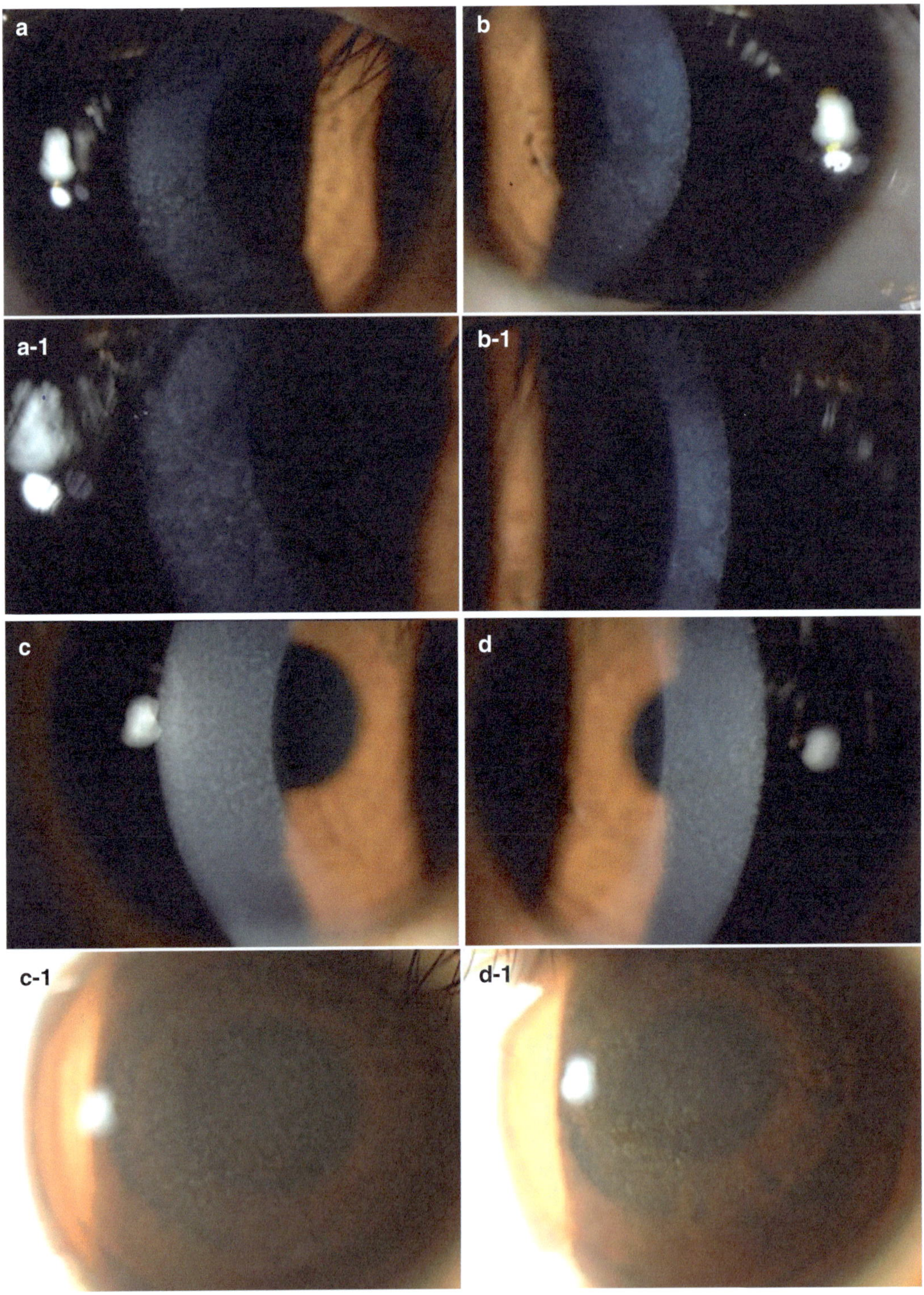

Fig. 3.4 Anterior segment photographs of case 3. A 14-year-old female presented with corneal opacities and recurrent ocular pain in both eyes (**a, a-1, b, b-1**). High magnification view of the (**a**) and (**b**) showed partially dropped off corneal deposits in the center of each cornea (**a-1, b-1**). Seven years later, typical honeycomb-like corneal opacities were observed in both eyes (**c, c-1, d, d-1**). The opacities did not reach the limbus. The eyes were no longer painful. Best-corrected visual acuity was 0.7 in the right eye and 1.0 in the left eye

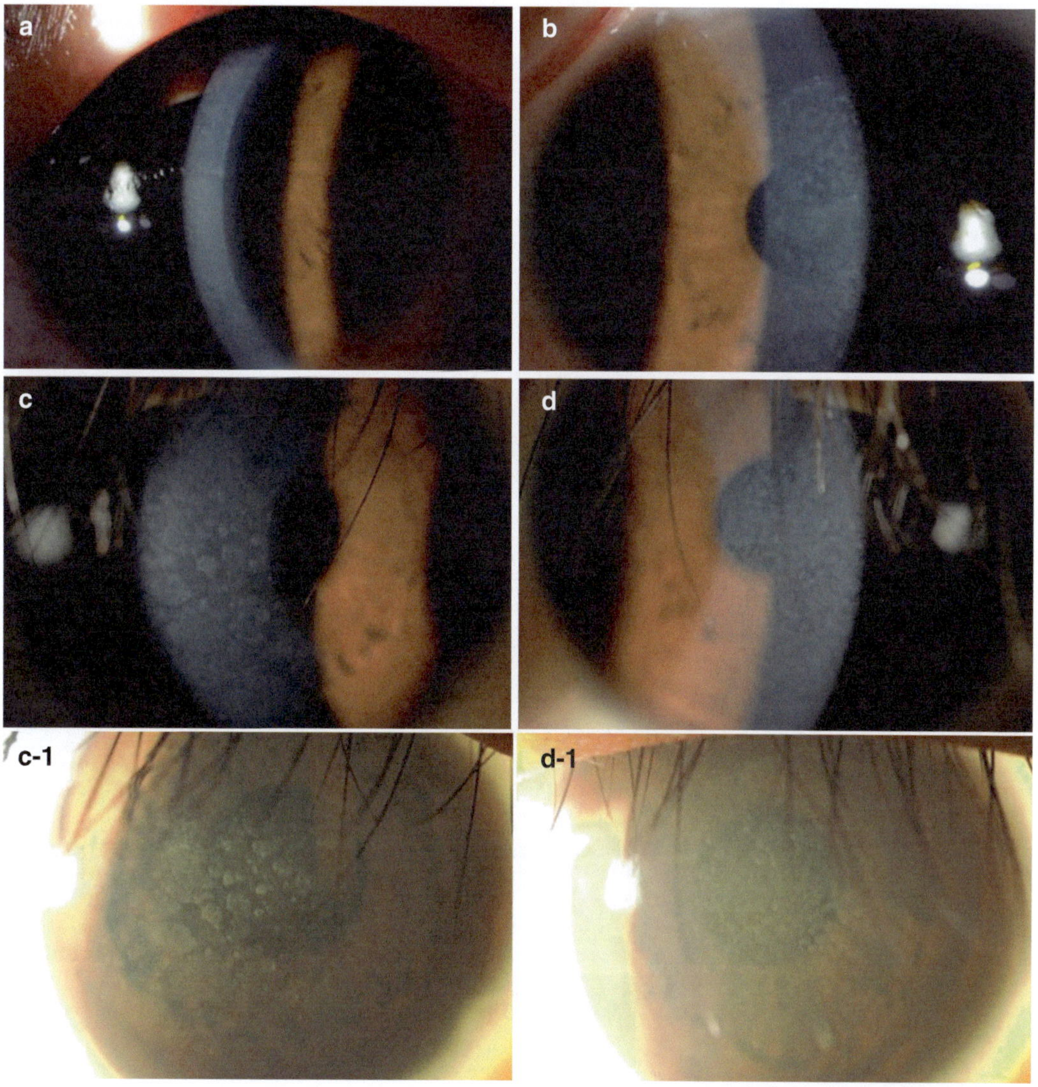

Fig. 3.5 Anterior segment photographs of case 4. Right eye (**a, c, c-1**), left eye (**b, d, d-1**). A 10-year-old female presented after insertion of therapeutic bandage contact lens in both eyes at another hospital because of corneal erosion (**a, b**; At presentation). Her uncorrected visual acuity was 0.6 in the right eye and 0.8 in the left eye. Seven years later, honeycomb-like corneal opacities with clear intervening stroma were observed in both eyes (**c, c-1, d, d-1**). Best-corrected visual acuity was 0.7 in the right eye and 0.9 in the left eye

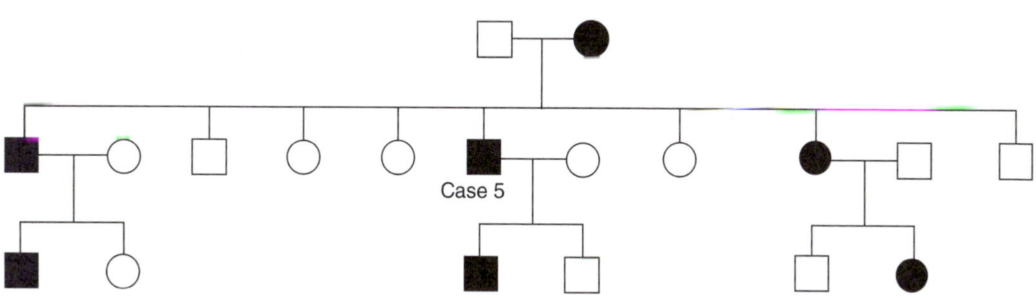

Fig. 3.6 A pedigree of case 5 affected by Thiel-Behnke corneal dystrophy

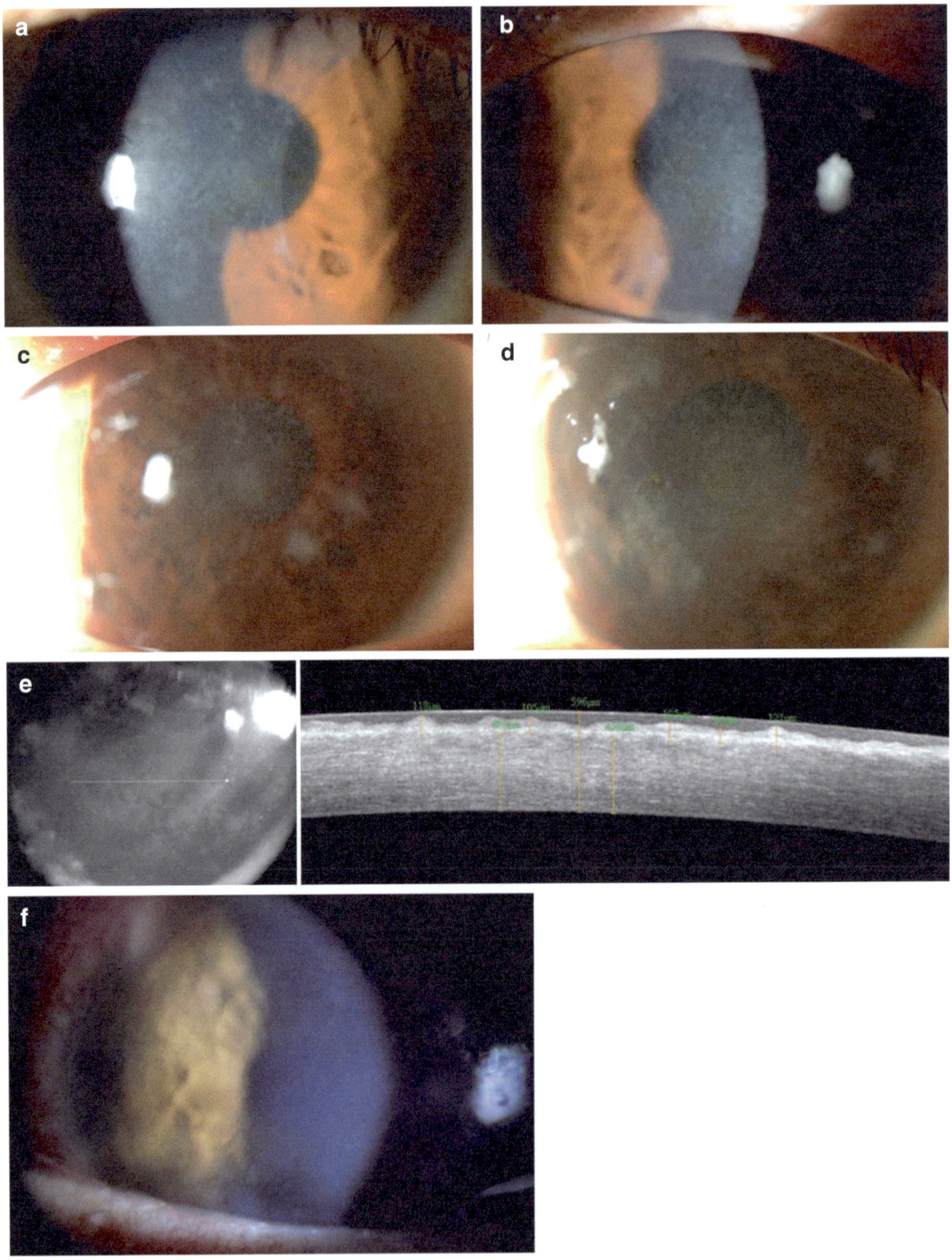

Fig. 3.7 Anterior segment photographs and anterior segment optical coherence tomography (AS-OCT) of case 5. Right eye (**a, c**), left eye (**b, d, e, f**). A 43-year-old male presented with decreased visual acuity in both eyes. At presentation, diffuse gray-whitish opacities and a brown iron line were observed in both eyes (**a–d**). Uncorrected and corrected visual acuities were 0.3 in both eyes. The thickness of corneal opacities ranged from 86 to 121 μm in the preoperative AS-OCT image (**e**). Phototherapeutic keratectomy (PTK) (80 μm-depth ablation) was performed in his left eye. Four days after PTK (**f**), opacities were no longer visible. The visual acuity was 0.4 in the left eye 1 month postoperatively

Clinical Manifestations

Fleck-like or punctate opacities appear in the center of cornea in early childhood. Patients experience painful, bilateral recurrent erosions and progressive deterioration of vision. The severity and frequency of erosive attacks are less than those of RBCD (Kuchle et al. 1995; Weiss et al. 2008). In the first and second decades, the clinical appearance of corneal deposits may be confused with that of RBCD because of depth and location of deposits. In the later stage, corneal deposits show a typical honeycomb appearance, which can be differentiated from the geographic appearance of opacities that are characteristic of RBCD. The corneal opacities reach limbus and deeper stroma with time. Iron lines in corneas with honeycomb-shape opacities have been observed in several patients (Kuchle et al. 1995) and are also seen in Figs. 3.2a, b, 3.3a, and 3.7a–d. These iron lines are caused by the accumulation of tear fluid around the epithelial irregularities—the same mechanism that causes other iron lines (Barraquer-Somers et al. 1983).

Anterior Segment Optical Coherence Tomography (AS-OCT) and Confocal Microscope Imaging

AS-OCT reveals hyper-reflective sawtooth-like deposits along Bowman layer (Nowinska et al. 2015; Yu et al. 2015), which have been considered by some authors to be a feature that distinguishes TBCD from the band-like homogenous deposits of RBCD (Liang et al. 2014; Qiu et al. 2016; Weiss et al. 2015). Other authors, however, have reported the sawtooth-like hyper-reflective deposits in RBCD patients (Fogla and Knyazer 2014; Liang et al. 2014; Weiss et al. 2015). Sawtooth-like AS-OCT findings, therefore, are not a feather that is pathognomonic for TBCD.

Confocal microscopy reveals focal reflective deposits in the superficial epithelium, homogenous reflective deposits in the epithelial basal layer, and irregular highly reflective material in the Bowman layer without alteration of keratocyte nuclei (Chen et al. 2010; Kobayashi et al. 2003; Kobayashi and Sugiyama 2007; Werner et al. 1999).

Microscopic Findings

Light microscopy reveals irregular homogenous deposits throughout Bowman layer, some of which cause epithelial irregularity (Kuchle et al. 1995). This fibrocellular tissue, which is located between the epithelium and stroma, has a sawtooth appearance (Lisch and Weiss 2019). Focal iron deposits are also observed in the basal epithelial layer in some patients. Empty vacuoles of variables sizes were present in the deep stroma, and there were no abnormalities in Descemet membrane and endothelium. (Kuchle et al. 1995).

Transmission electron microscopy shows curly fiber-shaped reflective deposits in Bowman layer in TBCD, which are distinct from the rod-shaped deposits of RBCD. The average diameter of a curly fiber is 9–15 nm (Kuchle et al. 1995). The Bowman layer is almost absent.

Treatments

Because the visually significant deposits are restricted to the anterior stroma, phototherapeutic keratectomy (PTK) is frequently the best, and least aggressive, surgical option for the treatment of patients with TBCD (Ellies et al. 2003; Fagerholm 2003; Seitz and Lisch 2011). Mean ablation depth from epithelium to stroma has been reported as 116 μm (range: 100–140 μm) (Hieda et al. 2013). Half of ten eyes of five patients showed recurrence with central subepithelial opacity during the average 60 months of follow-up (range: 12–108 months). Among the five eyes, the visual acuity decreased more than two lines in four eyes and one line in one eye. Lamellar keratoplasty (Kuchle et al. 1995) and penetrating keratoplasty (Ellies et al. 2003) are other therapeutic options, but recurrences are observed by these more aggressive treatment options, as well. Recently, PTK followed by wavefront-guided photorefractive keratectomy and topography-guided transepithelial PTK have been reported to reduce irregular corneal astigmatism and subjective visual aberrations (Baenninger and Reichmuth 2014; Hsiao and Hou 2017).

References

Baenninger PB, Reichmuth V. Topography-guided transepithelial photorefractive keratektomy (cTEN) for treatment of Thiel-Behnke dystrophy. Klin Monbl Augenheilkd. 2014;231:329–30.

Barraquer-Somers E, Chan CC, Green WR. Corneal epithelial iron deposition. Ophthalmology. 1983;90:729–34.

Chen YJ, Chen JT, Lu DW, Tai MC. In vivo corneal confocal microscopic findings and gene analysis of three patients with Thiel-Behnke corneal dystrophy. Br J Ophthalmol. 2010;94:262–4.

Ellies P, Bejjani RA, Bourges JL, Boelle PY, Renard G, Dighiero P. Phototherapeutic keratectomy for BIGH3-linked corneal dystrophy recurring after penetrating keratoplasty. Ophthalmology. 2003;110:1119–25.

Fagerholm P. Phototherapeutic keratectomy: 12 years of experience. Acta Ophthalmol Scand. 2003;81:19–32.

Fogla R, Knyazer B. Microkeratome-assisted two-stage technique of superficial anterior lamellar keratoplasty for Reis-Bücklers corneal dystrophy. Cornea. 2014;33:1118–22.

Hieda O, Kawasaki S, Wakimasu K, Yamasaki K, Inatomi T, Kinoshita S. Clinical outcomes of phototherapeutic keratectomy in eyes with Thiel-Behnke corneal dystrophy. Am J Ophthalmol. 2013;155:66–72.e61.

Hsiao CC, Hou YC. Combination of phototherapeutic keratectomy and wavefront-guided photorefractive keratectomy for the treatment of Thiel-Behnke corneal dystrophy. Indian J Ophthalmol. 2017;65:318–20.

Kobayashi A, Sugiyama K. In vivo laser confocal microscopy findings for Bowman's layer dystrophies (Thiel-Behnke and Reis-Bücklers corneal dystrophies). Ophthalmology. 2007;114:69–75.

Kobayashi A, Sakurai M, Shirao Y, Sugiyama K, Ohta T, Amaya-Ohkura Y. In vivo confocal microscopy and genotyping of a family with Thiel-Behnke (honeycomb) corneal dystrophy. Arch Ophthalmol. 2003;121:1498–9.

Kuchle M, Green WR, Volcker HE, Barraquer J. Reevaluation of corneal dystrophies of Bowman's layer and the anterior stroma (Reis-Bücklers and Thiel-Behnke types): a light and electron microscopic study of eight corneas and a review of the literature. Cornea. 1995;14:333–54.

Liang Q, Pan Z, Sun X, Baudouin C, Labbe A. Reis-Bücklers corneal dystrophy: a reappraisal using in vivo and ex vivo imaging techniques. Ophthalmic Res. 2014;51:187–95.

Lisch W, Kivela T. Individual phenotypic variances in a family with Thiel-Behnke corneal dystrophy. Cornea. 2013;32:e192–3.

Lisch W, Weiss JS. Clinical and genetic update of corneal dystrophies. Exp Eye Res. 2019;186:107715.

Nowinska AK, Teper SJ, Janiszewska DA, Lyssek-Boron A, Dobrowolski D, Koprowski R, Wylegala E. Comparative study of anterior eye segment measurements with spectral swept-source and time-domain optical coherence tomography in eyes with corneal dystrophies. Biomed Res Int. 2015;2015:805367.

Qiu WY, Zheng LB, Pan F, Wang BB, Yao YF. New histopathologic and ultrastructural findings in Reis-Bücklers corneal dystrophy caused by the Arg124Leu mutation of TGFBI gene. BMC Ophthalmol. 2016;16:158.

Seitz B, Lisch W. Stage-related therapy of corneal dystrophies. Dev Ophthalmol. 2011;48:116–53.

Sorour HM, Yee SB, Peterson NJ, Li FT, Macsai MS, Zhao XC, Yee RW. Recurrence of chromosome 10 Thiel-Behnke corneal dystrophy (CDB2) after excimer laser phototherapeutic keratectomy or penetrating keratoplasty. Cornea. 2005;24:45–50.

Sullivan LS, Zhao X, Bowne SJ, Xu X, Daiger SP, Yee SB, Yee RW. Exclusion of the human collagen type XVII (COL17A1) gene as the cause of Thiel-Behnke corneal dystrophy (CDB2) on chromosome 10q23-q25. Curr Eye Res. 2003;27:223–6.

Thiel HJ, Behnke H. A hitherto unknown subepithelial hereditary corneal dystrophy. Klin Monbl Augenheilkd. 1967;150:862–74.

Weiss JS, Moller HU, Lisch W, Kinoshita S, Aldave AJ, Belin MW, Kivela T, Busin M, Munier FL, Seitz B, Sutphin J, Bredrup C, Mannis MJ, Rapuano CJ, Van Rij G, Kim EK, Klintworth GK. The IC3D classification of the corneal dystrophies. Cornea. 2008;27 Suppl 2:S1–83.

Weiss JS, Moller HU, Aldave AJ, Seitz B, Bredrup C, Kivela T, Munier FL, Rapuano CJ, Nischal KK, Kim EK, Sutphin J, Busin M, Labbe A, Kenyon KR, Kinoshita S, Lisch W. IC3D classification of corneal dystrophies—edition 2. Cornea. 2015;34:117–59.

Werner LP, Werner L, Dighiero P, Legeais JM, Renard G. Confocal microscopy in Bowman and stromal corneal dystrophies. Ophthalmology. 1999;106:1697–704.

Yee RW, Sullivan LS, Lai HT, Stock EL, Lu Y, Khan MN, Blanton SH, Daiger SP. Linkage mapping of Thiel-Behnke corneal dystrophy (CDB2) to chromosome 10q23-q24. Genomics. 1997;46:152–4.

Yu Y, Qiu P, Zhu Y, Li J, Wu M, Zhang B, Yao K. A novel phenotype-genotype correlation with an Arg555Trp mutation of TGFBI gene in Thiel-Behnke corneal dystrophy in a Chinese pedigree. BMC Ophthalmol. 2015;15:131.

TGFBI Corneal Dystrophies: Lattice Corneal Dystrophy with *TGFBI* Mutation—Classic Lattice Corneal Dystrophy (Lattice Corneal Dystrophy Type 1)

4

Kyung Eun Han, Roo Min Jun, Sun Woong Kim, and R. Doyle Stulting

Classic lattice corneal dystrophy (classic LCD, lattice corneal dystrophy type 1) is an autosomal dominantly inherited epithelial-stromal corneal dystrophy associated with an R124C (Arg124Cys) mutation in the transforming growth factor beta-induced (*TGFBI*) gene on chromosome 5q31. Lattice dystrophy of the cornea was first described by Biber (1890) and by Haab (1899) and Dimmer (1899). Classic LCD has also been known as Biber-Haab-Dimmer dystrophy.

K. E. Han (✉)
Department of Ophthalmology, Ewha Womans University Mokdong Hospital, Ewha Womans University College of Medicine, Seoul, Korea (Republic of)
e-mail: hanke@ewha.ac.kr

R. M. Jun
Department of Ophthalmology, Ewha Womans University Seoul Hospital, Ewha Womans University College of Medicine, Seoul, Korea (Republic of)
e-mail: jrmoph@ewha.ac.kr

S. W. Kim
Department of Ophthalmology, Yonsei University Wonju Severance Christian Hospital, Yonsei University Wonju College of Medicine, Wonju, Korea (Republic of)
e-mail: eyedockim@yonsei.ac.kr

R. D. Stulting
Department of Ophthalmology, Emory University, Atlanta, GA, USA

Woolfson Eye Institute, Atlanta, GA, USA
e-mail: dstulting@icloud.com

Clinical Manifestation

Classic LCD is characterized by freckle-like subepithelial opacities in the center of the cornea that appear during the first decade of life. Fine branching and non-branching refractile opacities are best visualized in the center and mid-periphery of the cornea by retroillumination (Fig. 4.1b). These lines elongate and enlarge, spreading from the central into the peripheral cornea and involving the deeper corneal stroma during the second and third decade, but they do not reach the far peripheral cornea (Fig. 4.2a, b). Descemet membrane and endothelium are uninvolved.

With age, the central dots and lattice lines become confluent and are accompanied by recurrent corneal erosions. Ground-glass haze appears in the central cornea, reducing visual acuity by the fourth decade (Figs. 4.3a and 4.4b).

Classic LCD must be differentiated from LCD variants including LCD IIIA, I/IIIA, IV, and atypical forms. Phenotypes of LCD variants vary from a pattern of large, ridged round peripheral opacities with uneven surfaces and thin lattice lines (Tian et al. 2007) to that of an irregular surface with diffuse whitish-gray

E. K. Kim, R. D. Stulting (eds.), *TGFBI-related Corneal Dystrophies*,
https://doi.org/10.1007/978-981-96-0131-8_4

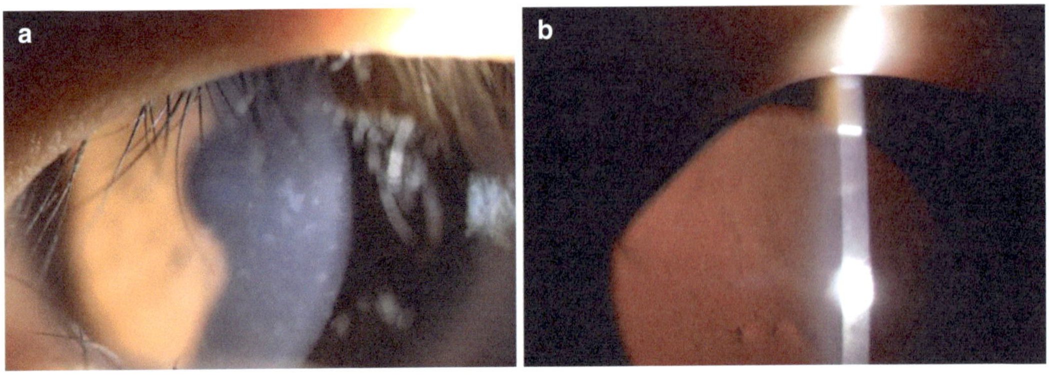

Fig. 4.1 Slit lamp photographs of classic lattice corneal dystrophy under direct (**a**) and indirect retroillumination (**b**) of in a 13-year-old male patient. Multiple, gray-whitish freckle-like opacities are in the center and midperipheral cornea (**a**). In retroillumination, radiating refractile lattice lines can be seen (**b**)

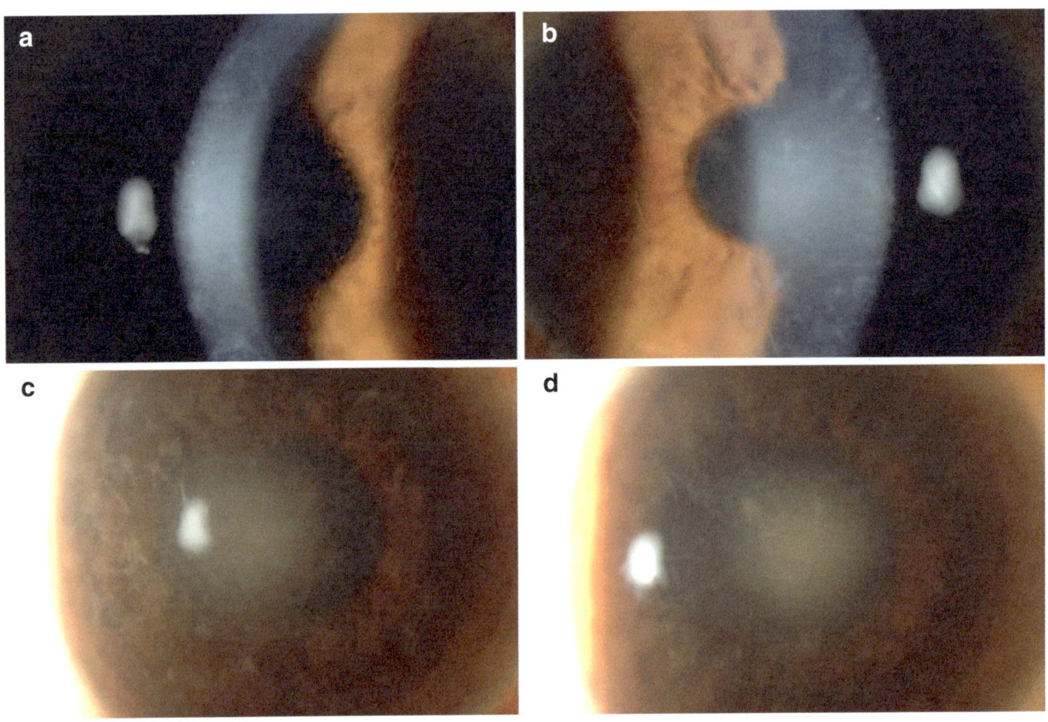

Fig. 4.2 Anterior segment photographs of a 44-year-old male with classic lattice corneal dystrophy. (**a, c**) Right eye, (**b, d**) left eye. In both eyes, there is central corneal stromal clouding. Fine, refractile branching lattice lines and freckle-like patches of subepithelial haze can be seen around the area of central clouding

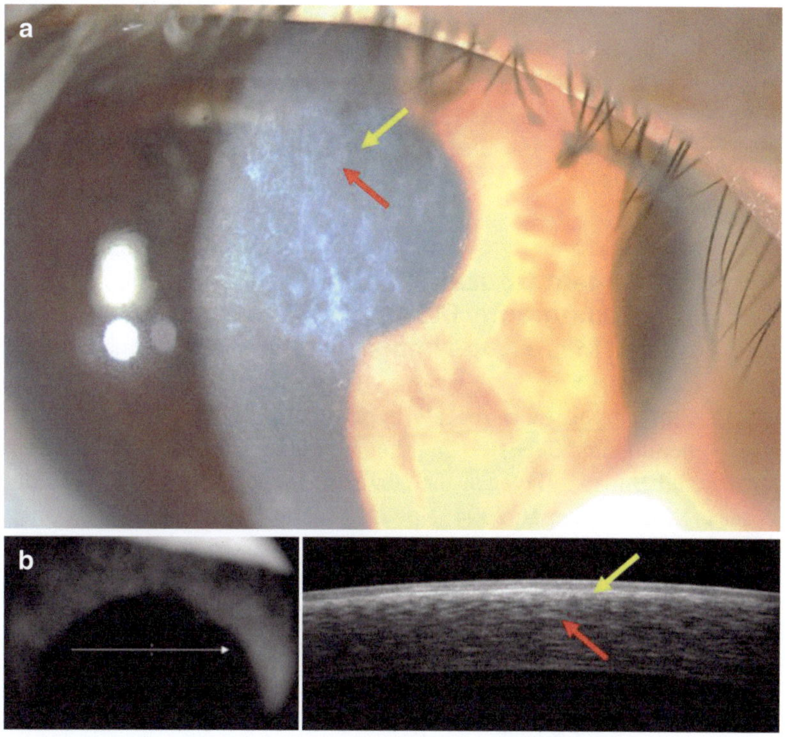

Fig. 4.3 Anterior segment photograph (**a**) and anterior segment optical coherence tomography (**b**) of the right eye of a 25-year-old female with classic lattice corneal dystrophy. Uncorrected visual acuity was 0.5 in the right eye and 0.4 in the left eye. Mild subepithelial haze is seen in the central and midperipheral cornea (yellow arrow), and thick lattice lines are seen in the deep stroma (red arrow). This patient's mother required penetrating keratoplasty at the age of 50

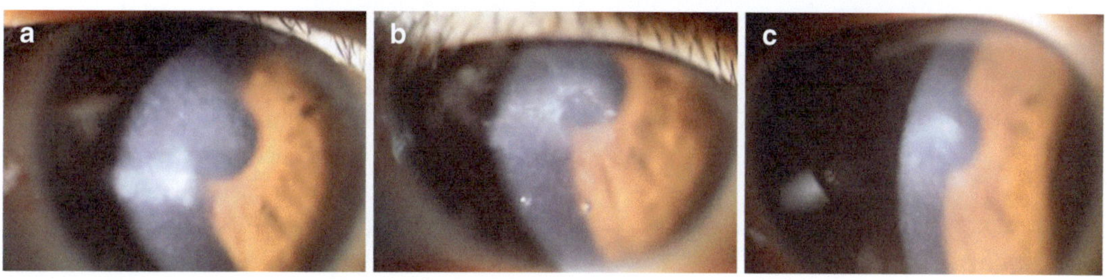

Fig. 4.4 Anterior segment photographs of the right eye of the same patient shown in Fig. 4.3 at the age of 37. The opacities merged to form a ground-glass opacity at the center of the cornea, reducing her vision to 0.2 (**a**). She underwent a 50 μm phototherapeutic keratectomy (PTK) that successfully removed most of the central opacities; however, the central epithelial surface remained irregular 1 month after treatment (**b**). Visual acuity improved to 0.5 20 months after PTK (**c**)

deposits in the anterior stroma, along with refractile lines radially spreading to the peripheral cornea (Ji et al. 2022). Until now, more than ten mutations are reported as LCD variants and even more are being found (Han et al. 2016; Weiss et al. 2024).

Anterior Segment Optical Coherence Tomography (AS-OCT) and Confocal Microscope Imaging

AS-OCT of classic LCD reveals hyperreflective subepithelial deposits in the anterior stroma and Bowman layer at the center of the cornea (Klintworth et al. 2004; Nowinska et al. 2011; Fig. 4.3b). Other authors reported an increased reflectivity corresponding to corneal opacities with a distinct border between the corneal opacity and normal cornea (Nowinska et al. 2011).

Confocal microscopy reveals that opacities as bright deposits (Klintworth et al. 2004) or highly reflective branching filaments (Kobayashi et al. 2007) in the superficial and mid-stroma. Highly reflective reticular extracellular deposits are also seen in the basal epithelium (Kobayashi et al. 2007).

Microscopic Findings

Light microscopy reveals stromal deposits that are positive for amyloid under direct fast scarlet stain and orange with Congo red stain. These deposits show red-green birefringence under polarized light. Deposits in the anterior stroma distort the structure of the corneal lamellae. Focal absence of Bowman layer and atrophy of the epithelium are noticeable (Dighiero et al. 2001; Kobayashi et al. 2007; Nowinska et al. 2011).

By transmission electron microscopy, fine, electron-dense extracellular fibrils of 8–10 nm diameter are observed (Dighiero et al. 2001; Resch et al. 2009). These amyloid fibrils are intermingled with electron-lucent and electron-dense cellular debris between basal epithelial cells and Bowman layer (Resch et al. 2009).

Treatment

When central opacities are dense enough to reduce visual acuity, but do not affect the deep stroma, phototherapeutic keratectomy (PTK) may be considered. In advanced stages, deep anterior lamellar keratoplasty or penetrating keratoplasty may be required to restore functional vision.

PTK may successfully remove anterior stromal deposits, but it does not prevent recurrences. Chen and Xie reported recurrences in ten eyes of six patients with LCD (including classic LCD and variants) an average of 19.7 months (range, 6–46 months) after PTK with a depth of 114.8 ± 16.6 μm (Chen and Xie 2013). Notably, corneal flattening produced by PTK creates a hyperopic shift (1.0 ± 1.8 D, median: 0.5 D) (Das et al. 2005b).

Because the recurrences after PTK are superficial, PTK can be repeated, as long as the thickness of residual stromal bed is expected to be more than about 250 μm.

After PKP, recurrences appear as subepithelial opacities, anterior stromal haze, or both after about 9 years (range, 3–26 years) (Meisler and Fine 1984). A second procedure was needed in half of 20 eyes from 14 patients, after a mean of 11.3 years (range, 9.3–14.0 years) when there was a visually significant recurrence, defined as visual loss of more than four lines of best-corrected visual acuity (BCVA) and/or BCVA ≤20/80 (Ellies et al. 2002).

After PTK or keratoplasty, delayed epithelial wound healing (>7 days) is often seen (Das et al. 2005a, b; Foerster et al. 2007), as also observed in Fig. 4.4a–c. Delayed epithelialization may be attributed to amyloid deposition itself or alterations in the basement membrane (Foerster et al. 2007). These patients should be followed closely until epithelialization is complete (Figs. 4.5, 4.6, and 4.7).

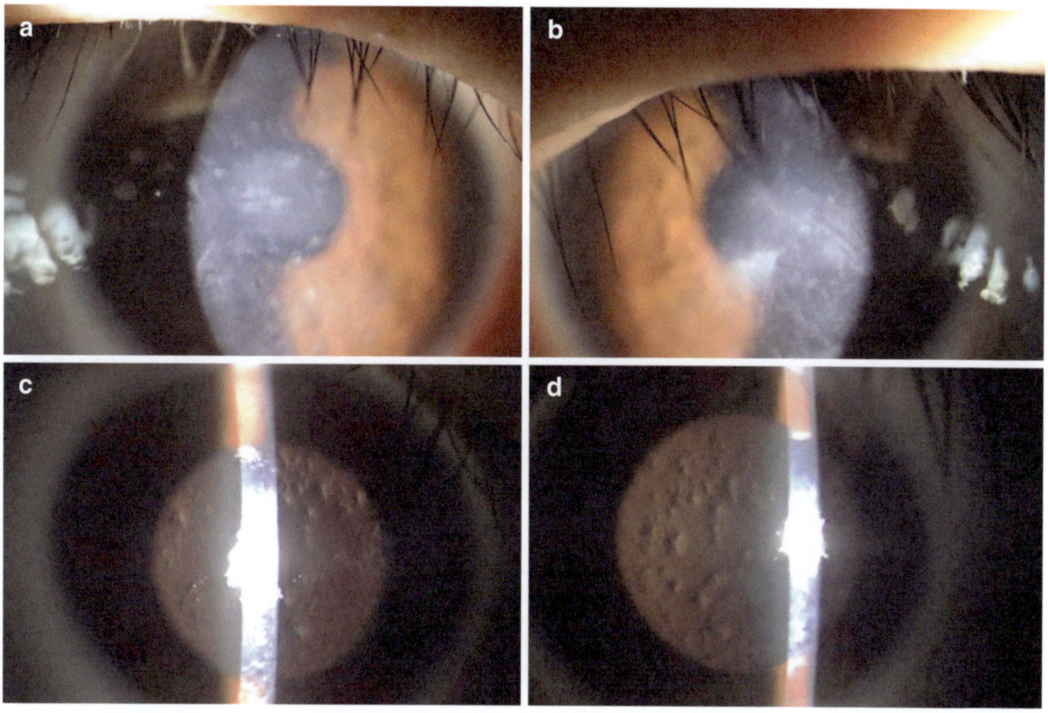

Fig. 4.5 Anterior segment photographs of the right eye (**a, c**) and the left eye (**b, d**) under direct (**a, b**) and indirect (**c, d**) retroillumination of a 31-year-old female with classic lattice corneal dystrophy. Best-corrected visual acuity was 0.2 in the right eye and 0.15 in the left eye. In both eyes, there were diffuse, dot-shaped opacities and circular ground-glass opacities combined with thick merged lattice lines in the pupillary area. Thin branching and non-branching lines are seen in the mid-peripheral cornea

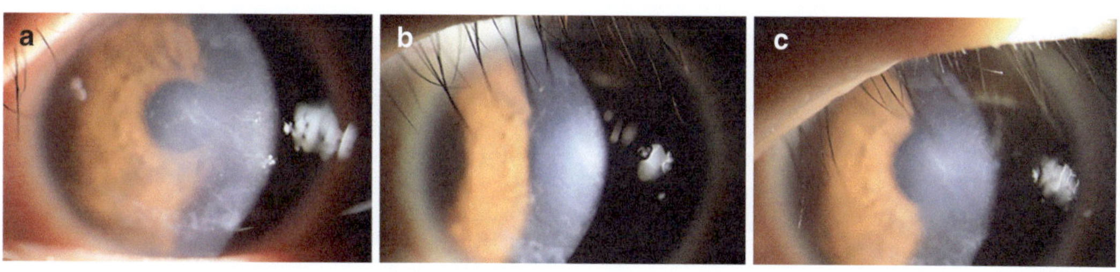

Fig. 4.6 Postoperative anterior segment photographs of the left eye of the patient seen in Fig. 4.5 who underwent a 20 μm phototherapeutic keratectomy (PTK). At postoperative day 3, the central opacity was almost completely removed, but some deep lattice lines remained. Best-corrected visual acuity improved to 0.4 at postoperative day 10 (**a**). At 2 months postoperatively, the intraocular pressure of the left eye rose to 25 mmHg due to topical steroids, and central cloudiness related to corneal edema developed (**b**). A combination of timolol and dorzolamide eye drops stabilized intraocular pressure, and by 5 months postoperatively, the central corneal haze had been minimized (**c**)

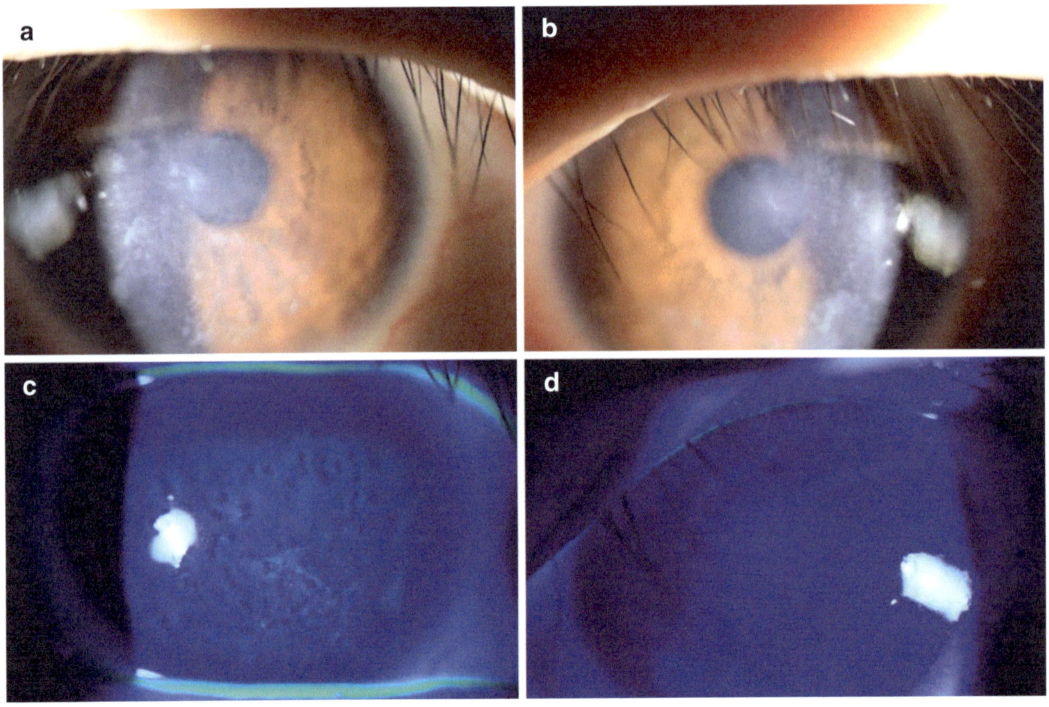

Fig. 4.7 Anterior segment photographs of the right eye (**a, c**) and the left eye (**b, d**) in direct illumination (**a, b**) and after fluorescein stain (**c, d**) of the patient seen in Fig. 4.5. One year postoperatively, best-corrected visual acuity had improved to 0.3 in the right eye and 0.4 in the left eye. The corneal opacity and surface irregularity of the right cornea, which did not undergo PTK, remained similar to that seen in Fig. 4.5a. The left eye, which underwent PTK, showed a reduction in the density of central corneal haze and smoothing of the surface

References

Biber H. Ober einige seltenere hornhauterkrankungen: die oberflichliche gittrige keratitis. Inaugural Dissertation, A. Diggelmann: Zurich; 1890. p. 35–42.

Chen M, Xie L. Features of recurrence after excimer laser phototherapeutic keratectomy for anterior corneal pathologies in North China. Ophthalmology. 2013;120:1179–85.

Das S, Langenbucher A, Seitz B. Delayed healing of corneal epithelium after phototherapeutic keratectomy for lattice dystrophy. Cornea. 2005a;24:283–7.

Das S, Langenbucher A, Seitz B. Excimer laser phototherapeutic keratectomy for granular and lattice corneal dystrophy: a comparative study. J Refract Surg. 2005b;21:727–31.

Dighiero P, Niel F, Ellies P, D'Hermies F, Savoldelli M, Renard G, Delpech M, Valleix S. Histologic phenotype-genotype correlation of corneal dystrophies associated with eight distinct mutations in the TGFBI gene. Ophthalmology. 2001;108:818–23.

Dimmer F. Ober oberflichliche gitterige hornhauttrubung. Zeitschrift fir Augenheilkunde. 1899;2:354–61.

Ellies P, Renard G, Valleix S, Boelle PY, Dighiero P. Clinical outcome of eight BIGH3-linked corneal dystrophies. Ophthalmology. 2002;109:793–7.

Foerster CG, Langenbucher A, Cursiefen C, Kruse FE, Seitz B. Delayed epithelial healing after keratoplasty for lattice corneal dystrophy. Cornea. 2007;26:1182–3.

Haab O. Die gittrige keratitis. Zeitschrift fur Augenheilkunde. 1899;2:235–7.

Han KE, Choi SI, Kim TI, Maeng YS, Stulting RD, Ji YW, Kim EK. Pathogenesis and treatments of TGFBI corneal dystrophies. Progr Retinal Eye Res. 2016;50:67–88.

Ji YW, Ahn H, Shin KJ, Kim TI, Seo KY, Stulting RD, Kim EK. De Novo L509P mutation of the TGFBI gene associated with slit-lamp findings of lattice corneal dystrophy type IIIA. J Clin Med. 2022;11:3055.

Klintworth GK, Bao W, Afshari NA. Two mutations in the TGFBI (BIGH3) gene associated with lattice corneal dystrophy in an extensively studied family. Invest Ophthalmol Vis Sci. 2004;45:1382–8.

Kobayashi A, Fujiki K, Fujimaki T, Murakami A, Sugiyama K. In vivo laser confocal microscopic findings of corneal stromal dystrophies. Arch Ophthalmol. 2007;125:1168–73.

Meisler DM, Fine M. Recurrence of the clinical signs of lattice corneal dystrophy (type I) in corneal transplants. Am J Ophthalmol. 1984;97:210–4.

Nowinska AK, Wylegala E, Janiszewska DA, Dobrowolski D, Aragona P, Roszkowska AM, Puzzolo D. Genotype-phenotype correlation of TGFBI corneal dystrophies in Polish patients. Mol Vis. 2011;17:2333–42.

Resch MD, Schlotzer-Schrehardt U, Hofmann-Rummelt C, Kruse FE, Seitz B. Alterations of epithelial adhesion molecules and basement membrane components in lattice corneal dystrophy (LCD). Graefes Arch Clin Exp Ophthalmol. 2009;247:1081–8.

Tian X, Fujiki K, Zhang Y, Murakami A, Li Q, Kanai A, Wang W, Hao Y, Ma Z. A novel variant lattice corneal dystrophy caused by association of mutation (V625D) in TGFBI gene. Am J Ophthalmol. 2007;144:473–5.

Weiss JS, Rapuano CJ, Seitz B, Busin M, Kivelä TT, Bouheraoua N, Bredrup C, Nischal KK, Chawla H, Borderie V, Kenyon KR, Kim EK, Møller HU, Munier FL, Berger T, Lisch W, 2024. IC3D Classification of Corneal Dystrophies-Edition 3. Cornea 43. 466–27.

Variants of Lattice Corneal Dystrophy (LCD): LCD Type III, IIIA, and IV

5

Hideki Fukuoka

The cornea is the thin transparent tissue located directly in front of the iris and pupil of the eye, and it is the most transparent tissue in the human body. Numerous genes contribute to the attainment and maintenance of the specific properties of corneal tissue that allow vision (Kinoshita et al. 2001), and recent advances in molecular biology have led to a deeper understanding of corneal physiology and disease at the molecular level.

Lattice corneal dystrophy (LCD) is a rare form of corneal dystrophy characterized by stromal amyloid deposits comprised of comma-shaped overlapping dots and branching filaments that create a networked lattice effect that can lead to and loss of vision. Worldwide, lattice corneal dystrophy type 1 (LCD1; R124C; p.Arg124Cys) (Fig. 5.1) is the most common type of LCD and one of the epithelial-stromal types of LCD that is caused by mutations of the transforming growth factor beta-induced (*TGFBI*) gene located on chromosome 5 at locus 5q3, also known as transforming growth factor beta-inducible gene-H3 (βig-H3). Minor mutations of domain 4 of *TGFBI* other than LCD1 are classified as LCD variants (Weiss et al. 2008).

LCD variants LCD III, IIIA (Fig. 5.2), and IV can be classified on the basis of ocular findings, including the location of the stromal amyloid deposits, the specific characteristics of the networked lattice, and the age of onset. Compared to LCD1, both LCD III and IIIA are characterized by thick, ropy lattice lines that are radially oriented in the mid-stromal region that appear in patients who are at or over 50 years of age. Visual acuity in these patients does not decrease until later in life. It should be noted that another form of LCD with an autosomal recessive inheritance pattern is designated as LCD III. The gene responsible for LCD III remains unknown, and as Hida and associates pointed out (Hida et al. 1987), no genetic studies have been published. Yamamoto and associates (Yamamoto et al. 1998) reported that LCD IIIA has an autosomal dominant inheritance pattern and that the associated mutation is p.Pro501Thr. Congo red staining and immunohistochemistry show that amyloid and p.Pro501Thr-mutated protein in the corneal subepithelial and intrastromal area accumulate and colocalize in LCD IIIA patients (Kawasaki et al. 1999). They are also frequently associated with recurrent epithelial erosions. In some patients, the density of the lattice lines may increase and progress to involve the posterior and peripheral areas of the corneal stroma, thus causing diffuse corneal stromal haze. LCD IV is characterized by the late onset of deep stromal amyloid deposits just anterior to Descemet's membrane (Fig. 5.3), in contrast to the anterior stromal opacities seen in LCD1, III, and

H. Fukuoka (✉)
Department of Ophthalmology, Kyoto Prefectural University of Medicine, Kyoto, Japan
e-mail: hfukuoka@koto.kpu-m.ac.jp

E. K. Kim, R. D. Stulting (eds.), *TGFBI-related Corneal Dystrophies*,
https://doi.org/10.1007/978-981-96-0131-8_5

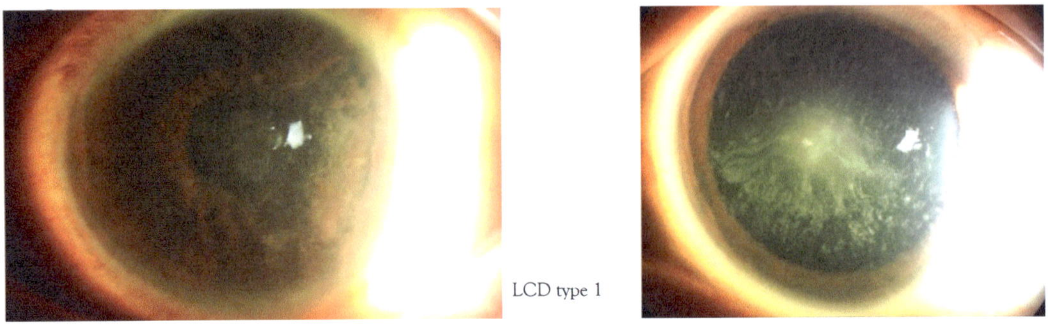

Fig. 5.1 Slit-lamp microscopy image of the eye in a typical case of lattice corneal dystrophy (LCD) type 1

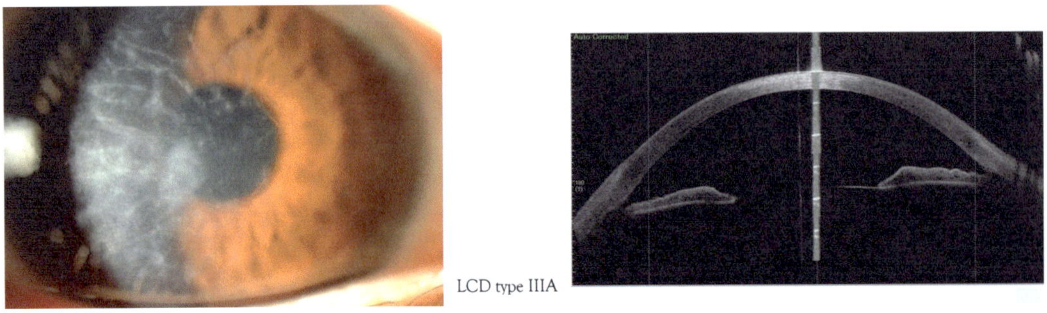

Fig. 5.2 Slit-lamp microscopy image of the eye in a typical case of LCD type IIIA (left-side image) (P501T). Anterior segment-optical coherence tomography (AS-OCT) image showing LCD IIIA deposits located in the middle and superficial stroma (right-side image)

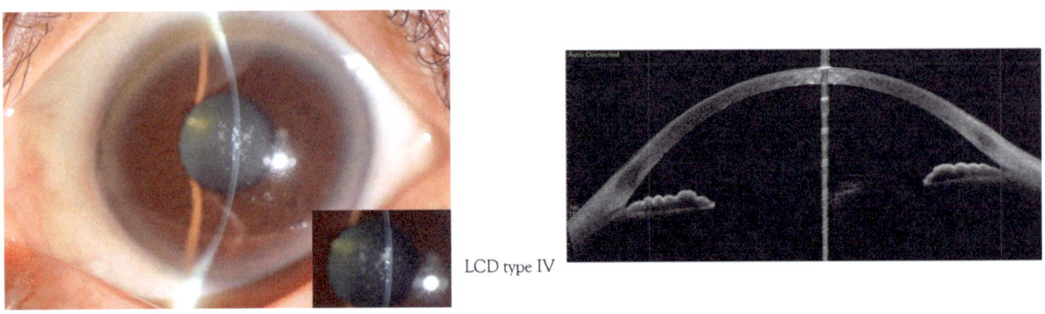

Fig. 5.3 Slit-lamp microscopy images of the eye in a typical case of LCD type IV (left-side image, inset) (L527R). AS-OCT image showing LCD type IV deposits located in the deep stroma (right-side image)

IIIA. With time, it progresses to involve more the more anterior corneal stroma. LCDIV, therefore, rarely cause recurrent epithelial erosions. The associated mutation is p.Leu527Arg (Fujiki et al. 1998). Although LCD1 has been reported worldwide, it is found primarily in the Japanese population. Thus, researchers have theorized that LCD IV may be caused by a founder mutation that occurred in a Japanese ancestor (Fukuoka et al. 2010).

It should be noted that numerous other mutations are reportedly associated with LCD variants. They include p.Arg496Trp (Kawasaki et al. 2011), p.Pro501Thr (Mashima et al. 2000; Yamamoto et al. 1998; Yu et al. 2006), p. Thr538Arg (Munier et al. 2002; Pampukha et al.

2008), p.Phe540del (Munier et al. 2002), p.Phe540Ser (Stix et al. 2005), p.Pro542Arg (Cho et al. 2012), p.Asn544Ser (Mashima et al. 2000; Yamada et al. 2009), p.Ala546Asp (Correa-Gomez et al. 2007; Eifrig et al. 2004), p.Ala546Thr (Dighiero et al. 2000; 2001), p.Phe547Ser (Takács et al. 2007), p.Leu558Pro (Livshits et al. 2008), p.Leu565Pro (Ołdak et al. 2014), p.His572del (Aldave et al. 2006), p.Gly594Val (Chakravarthi et al. 2005), p.Val613Gly (Niel-Butschi et al. 2011), p.Val613_Pro616del (Yang et al. 2010), p.Ala620Asp (Lakshminarayanan et al. 2011), p.Ala620Asp (Lakshminarayanan et al. 2011), p.Ala620Pro (Jung et al. 2014), p.Asn622His (Stewart et al. 1999), p.Asn622Lys (Munier et al. 2002), p.Gly623Asp (Munier et al. 2002), p.Val624Met (Afshari et al. 2008), p.His626Arg (Dighiero et al. 2001; Munier et al. 2002; Schmitt-Bernard et al. 2000; Stewart et al. 1999), p.His626Pro (Munier et al. 2002), p.Val624_Val625del (Chakravarthi et al. 2005), p.Val627Ter (Munier et al. 2002), p.Asn629_Pro630ins3 (Schmitt-Bernard et al. 2000), and p.Val631Asp (Munier et al. 2002), p.Val113Ile and Leu558Pro (Ann et al. 2017), p.Arg124Cys and Ala546Asp (Cao et al. 2017).

References

Afshari NA, Bahadur RP, Eifrig DE, Thogersen IB, Enghild JJ, Klintworth GK. Atypical asymmetric lattice corneal dystrophy associated with a novel homozygous mutation (Val624Met) in the TGFBI gene. MolE Vis. 2008;14:495–9.

Aldave AJ, Rayner SA, Kim BT, Prechanond A, Yellore VS. Unilateral lattice corneal dystrophy associated with the novel His572del mutation in the TGFBI gene. Mol Vis. 2006;12:142–6.

Ann LBT, Abbouda A, Frausto RF, Huseynli S, Gupta K, Alió JL, Aldave AJ. Variant lattice corneal dystrophy associated with compound heterozygous mutations in the TGFBI gene. Br J Ophthalmol. 2017;101(4):509–13. https://doi.org/10.1136/bjophthalmol-2015-307602.

Cao WP, Yuan HG, Liu P, Li X, Hu Q. Two mutations in the transforming growth factor beta-induced gene associated with familial lattice corneal dystrophy. Int J Ophthalmol. 2017;10(3):343–7. https://doi.org/10.18240/ijo.2017.03.03.

Chakravarthi SVVK, Kannabiran C, Sridhar MS, Vemuganti GK. TGFBI gene mutations causing lattice and granular corneal dystrophies in Indian patients. Investig Ophthalmol Vis Sci. 2005;46(1):121–5. https://doi.org/10.1167/iovs.04-0440.

Cho KJ, Mok JW, Na KS, Rho CR, Byun YS, Hwang HS, Hwang KY, Joo CK. TGFBI gene mutations in a Korean population with corneal dystrophy. Mol Vis. 2012;18:2012–21.

Correa-Gomez V, Villalvazo-Cordero L, Zenteno JC. The TGFBI A546D mutation causes an atypical type of lattice corneal dystrophy. Mol Vis. 2007;13:1695–700.

Dighiero P, Drunat S, Ellies P, D'Hermies F, Savoldelli M, Legeais JM, Renard G, Delpech M, Grateau G, Valleix S. A new mutation (A546T) of the βig-h3 gene responsible for a French lattice corneal dystrophy type IIIA. Am J Ophthalmol. 2000;129(2):248–51. https://doi.org/10.1016/S0002-9394(99)00324-4.

Dighiero P, Niel F, Ellies P, D'Hermies F, Savoldelli M, Renard G, Delpech M, Valleix S. Histologic phenotype-genotype correlation of corneal dystrophies associated with eight distinct mutations in the TGFBI gene. Ophthalmology. 2001;108(4):818–23. https://doi.org/10.1016/S0161-6420(00)00662-X.

Eifrig DE, Afshari NA, Buchanan HW IV, Bowling BL, Klintworth GK. Polymorphic corneal amyloidosis: a disorder due to a novel mutation in the transforming growth factor β-induced (BIGH3) gene. Ophthalmology. 2004;111(6):1108–14. https://doi.org/10.1016/j.ophtha.2003.09.043.

Fujiki K, Hotta Y, Nakayasu K, Yokoyama T, Takano T, Yamaguchi T, Kanai A. A new L527R mutation of the βIGH3 gene in patients with lattice corneal dystrophy with deep stromal opacities. Hum Genet. 1998;103(3):286–9. https://doi.org/10.1007/s004390050818.

Fukuoka H, Kawasaki S, Yamasaki K, Matsuda A, Fukumoto A, Murakami A, Kinoshita S. Lattice corneal dystrophy type iv (P.Leu527Arg) Is caused by a founder mutation of the TGFBI gene in a single Japanese ancestor. Investig Ophthalmol Vis Sci. 2010;51(9):4523–30. https://doi.org/10.1167/iovs.10-5343.

Hida T, Proia AD, Kigasawa K, Sanfilippo FP, Burchette JL, Akiya S, Klintworth GK. Histopathologic and immunochemical features of lattice corneal dystrophy type III. Am J Ophthalmol. 1987;104(3):249–54. https://doi.org/10.1016/0002-9394(87)90412-0.

Jung JW, Kim SA, Kang EM, Kim TI, Cho HS, Kim EK. Lattice corneal dystrophy type IIIA with hyaline component from a novel A620P mutation and distinct surgical treatments. Cornea. 2014;33(12):1324–31. https://doi.org/10.1097/ICO.0000000000000281.

Kawasaki S, Nishida K, Quantock AJ, Dota A, Bennett K, Kinoshita S. Amyloid and Pro501Thr-mutated βig-h3 gene product colocalize in lattice corneal dystrophy type IIIA. Am J Ophthalmol. 1999;127(4):456–8. https://doi.org/10.1016/S0002-9394(98)00360-2.

Kawasaki S, Yagi H, Yamasaki K, Matsuda A, Takeda K, Kinoshita S. A novel mutation of the TGFBI gene causing a lattice corneal dystrophy with deep stromal involvement. Br J Ophthalmol. 2011;95(1):150–1. https://doi.org/10.1136/bjo.2010.182667.

Kinoshita S, Adachi W, Sotozono C, Nishida K, Yokoi N, Quantock AJ, Okubo K. Characteristics of the human ocular surface epithelium. Prog Retin Eye Res. 2001;20(5):639–73. https://doi.org/10.1016/S1350-9462(01)00007-6.

Lakshminarayanan R, Vithana EN, Chai SM, Chaurasia SS, Saraswathi P, Venkatraman A, Rojare C, Venkataraman D, Tan D, Aung T, Beuerman RW, Mehta JS. A novel mutation in transforming growth factor-beta induced protein (TGFβIp) reveals secondary structure perturbation in lattice corneal dystrophy. Br J Ophthalmol. 2011;95(10):1457–62. https://doi.org/10.1136/bjophthalmol-2011-300651.

Livshits L, Pampukha VM, Tereshchenko FA, Drozhyna GI. Gene symbol: TGFBI. Disease: corneal dystrophy, lattice type. Hum Genet. 2008;124(3):296–7.

Mashima Y, Yamamoto S, Inoue Y, Yamada M, Konishi M, Watanabe H, Maeda N, Shimomura Y, Kinoshita S. Association of autosomal dominantly inherited corneal dystrophies with BIGH3 gene mutations in Japan. Am J Ophthalmol. 2000;130(4):516–7. https://doi.org/10.1016/S0002-9394(00)00571-7.

Munier FL, Frueh BE, Othenin-Girard P, Uffer S, Cousin P, Wang MX, Héon E, Black GCM, Blasi MA, Balestrazzi E, Lorenz B, Escoto R, Barraquer R, Hoeltzenbein M, Gloor B, Fossarello M, Singh AD, Arsenijevic Y, Zografos L, Schorderet DF. BIGH3 mutation spectrum in corneal dystrophies. Investig Ophthalmol Vis Sci. 2002;43(4):949–54.

Niel-Butschi F, Kantelip B, Iwaszkiewicz J, Zoete V, Boimard M, Delpech M, Bourges JL, Renard G, D'Hermies F, Pisella PJ, Hamel C, Delbosc B, Valleix S. Genotype-phenotype correlations of TGFBI p.Leu509Pro, p.Leu509Arg, p.Val613Gly, and the allelic association of p.Met502Val-p.Arg555Gln mutations. Mol Vis. 2011;17:1192–202.

Ołdak M, Szaflik JP, Ściezyńska A, Udziela M, Maksym RB, Rymgayłło-Jankowska B, Hofmann-Rummelt C, Menzel-Severing J, Płoski R, Zarnowski T, Kruse FE, Szaflik J. Late-onset lattice corneal dystrophy without typical lattice lines caused by a novel mutation in the TGFBI gene. Cornea. 2014;33(3):294–9. https://doi.org/10.1097/ICO.0000000000000062.

Pampukha VN, Kravchenko SA, Tereshchenko F, Drozhzhina GI, Livshits LA. TGFBI gene mutations in the Ukrainian patients with inherited corneal stromal dystrophies. Genetika. 2008;44(10):1392–6.

Schmitt-Bernard CF, Guittard C, Arnaud B, Demaille J, Argiles A, Claustres M, Tuffery-Giraud S. BIGH3 exon 14 mutations lead to intermediate type I/IIIA of lattice corneal dystrophies. Investig Ophthalmol Vis Sci. 2000;41(6):1302–8.

Stewart H, Black GCM, Donnai D, Bonshek RE, McCarthy J, Morgan S, Dixon MJ, Ridgway AAE. A mutation within exon 14 of the TGFBI (BIGH3) gene on chromosome 5q31 causes an asymmetric, late-onset form of lattice corneal dystrophy. Ophthalmology. 1999;106(5):964–70. https://doi.org/10.1016/S0161-6420(99)00539-4.

Stix B, Leber M, Bingemer P, Gross C, Rüschoff J, Fändrich M, Schorderet DF, Vorwerk CK, Zacharias M, Roessner A, Röcken C. Hereditary lattice corneal dystrophy is associated with corneal amyloid deposits enclosing C-terminal fragments of keratoepithelin. Investig Ophthalmol Vis Sci. 2005;46(4):1133–9. https://doi.org/10.1167/iovs.04-1319.

Takács L, Losonczy G, Matesz K, Balogh I, Sohajda Z, Tóth K, Fazakas F, Vereb G, Berta A. TGFBI (BIGH3) gene mutations in Hungary—report of the novel F547S mutation associated with polymorphic corneal amyloidosis. Mol Vis. 2007;13:1976–83.

Weiss JS, Møller HU, Lisch W, Kinoshita S, Busin M, Aldave AJ, Belin MW, Kivela T, Munier FL, Seitz B, Sutphin J, Bredrup C, Mannis MJ, Rapuano CJ, Van Rij G, Kim EK, Klintworth GK. The IC3D classification of the corneal dystrophies. Cornea. 2008;27:S1–S42. https://doi.org/10.1097/ico.0b013e31817780fb.

Yamada N, Kawamoto K, Morishige N, Chikama TI, Nishida T, Nishioka M, Okayama N, Hinoda Y. Double mutation (R124H, N544S) of TGFBI in two sisters with combined expression of Avellino and lattice corneal dystrophies. Mol Vis. 2009;15:974–9.

Yamamoto S, Okada M, Tsujikawa M, Shimomura Y, Nishida K, Inoue Y, Watanabe H, Maeda N, Kurahashi H, Kinoshita S, Nakamura Y, Tano Y. A keratoepithelin (βig-h3) mutation in lattice corneal dystrophy type IIIA [4]. Am J Hum Genet. 1998;62(3):719–22. https://doi.org/10.1086/301765.

Yang J, Han X, Huang D, Yu L, Zhu Y, Tong Y, Zhu B, Li C, Weng M, Ma X. Analysis of TGFBI gene mutations in Chinese patients with corneal dystrophies and review of the literature. Mol Vis. 2010;16:1186–93.

Yu P, Gu Y, Yang Y, Yan X, Chen L, Ge Z, Qi M, Si J, Guo L. A clinical and molecular-genetic analysis of Chinese patients with lattice corneal dystrophy and novel Thr538Pro mutation in the TGFBI (BIGH3) gene. J Genet. 2006;85(1):73–6. https://doi.org/10.1007/BF02728974.

Granular Corneal Dystrophy, Type 1 (Classic) (GCD1)

6

Hun Lee and R. Doyle Stulting

Clinical Features

Granular corneal dystrophy type 1 (GCD1), also known as Groenouw type I, is a bilateral hereditary corneal disorder characterized by the deposition of multiple, small, discrete, well-defined, grayish-white opacities in the anterior central stroma, sparing the limbus, during the first or second decade of life (Fig. 6.1) (Moller 1991; Weiss et al. 2015). The opacities of granular dystrophy may vary in shape and are classified into three basic morphologic types: drop-shaped, crumb-shaped, and ring-shaped (Weidle and Lisch 1984). The overall pattern of deposition is ray—or disk-shaped (Weidle and Lisch 1984). Initially, the stroma between the opacities remains clear. Although there is usually no associated discomfort or decrease in vision in the early stages of the disease, some patients may have mild photophobia from light scattering by corneal lesions. Glare, photophobia, and painful recurrent erosions often occur in advanced stages of the disease.

As the condition advances, individual lesions increase in size and number and may coalesce. They frequently extend into the deeper and more peripheral stroma. However, 2–3 mm of the peripheral cornea usually remains free of deposits (Fig. 6.1). With more advanced disease, the intervening cornea develops a diffuse, ground-glass appearance. In GCD1, many granular deposits are found not only in the superficial stroma but also between the basal cell layer of epithelium and Bowman's layer (Akhtar et al. 1999). The lesions of recurrent erosion can involve Bowman's layer and result in superficial irregularity.

Visual impairment is rare before the fifth decade and usually occurs secondary to the opacification of the intervening stroma. Corneal sensation is variably affected. Individuals who are homozygous for the mutant allele typically demonstrate an earlier onset and more severe manifestation of the phenotype. Three patients with a severe phenotype, proven to be homozygotes, were reported in a family from Japan (Okada et al. 1998). Clinical manifestations appeared at the early age of 6 years. Severe placoid corneal deposits developed later, and disease recurred soon after penetrating keratoplasty. Variants of granular dystrophy with an atypical appearance have also been associated with retinal findings consistent with those of cone dystrophy

H. Lee (✉)
Department of Ophthalmology, Asan Medical Center, University of Ulsan College of Medicine, Seoul, Korea (Republic of)
e-mail: yhun777@amc.seoul.kr

R. D. Stulting
Department of Ophthalmology, Emory University, Atlanta, GA, USA

Woolfson Eye Institute, Atlanta, GA, USA
e-mail: dstulting@icloud.com

E. K. Kim, R. D. Stulting (eds.), *TGFBI-related Corneal Dystrophies*,
https://doi.org/10.1007/978-981-96-0131-8_6

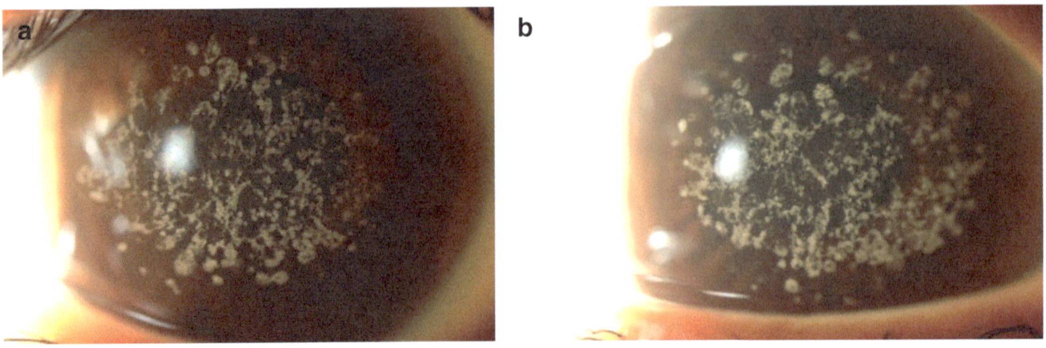

Fig. 6.1 Axially distributed discrete stromal opacities with intervening clear stroma and sparing of the periphery. Right eye (**a**) and left eye (**b**) of the same patient

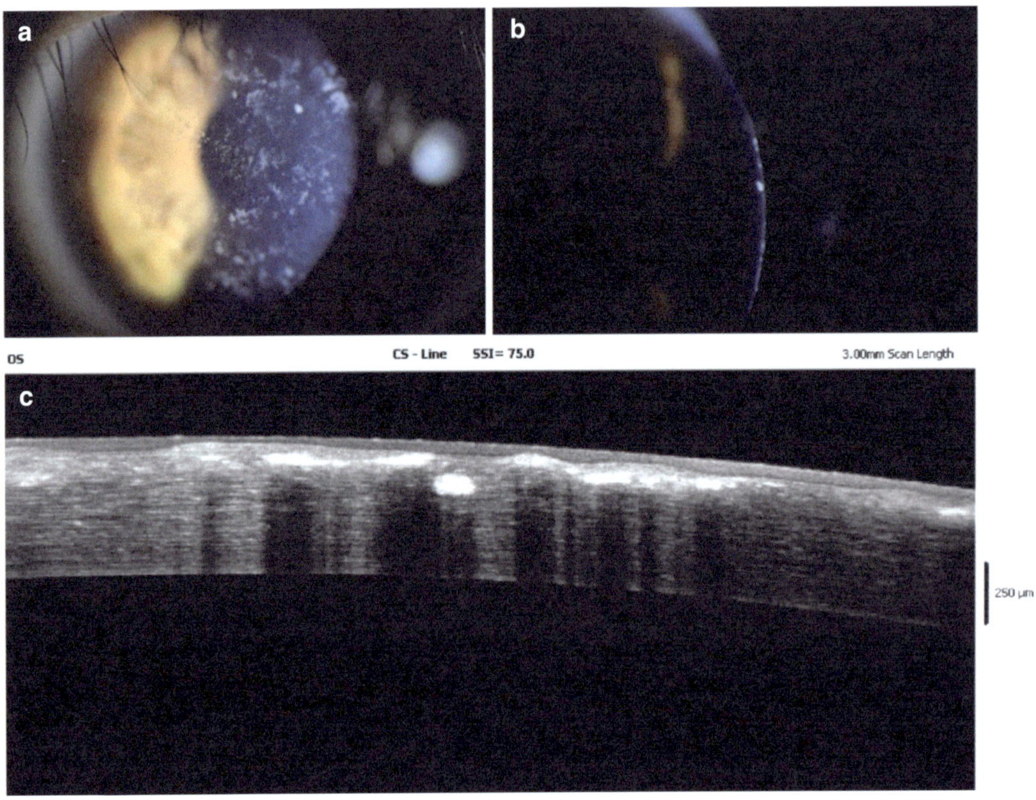

Fig. 6.2 Slit lamp photographs (**a, b**) and corneal OCT (**c**) demonstrating the shape of the lesions and their primarily anterior stromal location

(Haddad et al. 1977; Forsius et al. 1983; Rodrigues et al. 1984; Sekimoto et al. 1988).

Slit lamp examination in the early stages of the disease reveals fine dots, which appear opaque on direct illumination and may appear translucent on retroillumination, as well as radial lines in the superficial stroma. Typically, the lesions are randomly distributed and may be individual or aggregated into patterns (Fig. 6.2). The intervening stromal opacification that develops as the disease advances can best be seen as a slight haze with oblique illumination or as a discrete granu-

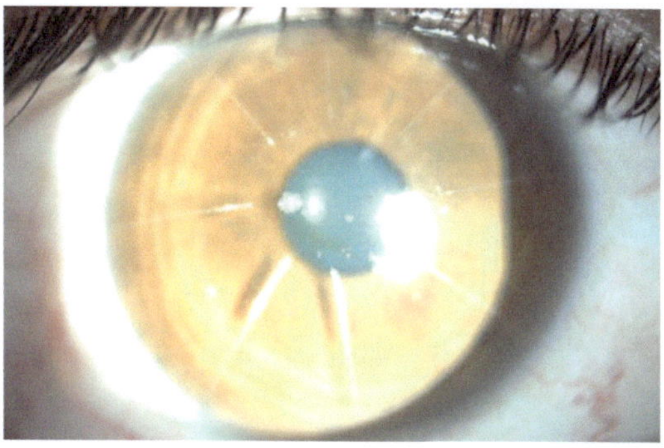

Fig. 6.3 Slit-lamp photograph reveals exacerbation of granular deposits in a GCD1 patient who underwent radial keratectomy (RK) for the correction of myopia. Deposits around the central portion of the incisions were thicker and denser than those around the peripheral portion of the incisions (Courtesy of Sepehr Feizi, MD, MSc Shaheed Beheshti University, Iran)

larity on retroillumination. The surface may stain negatively with fluorescein, and there may be areas of rapid tear film breakup over the superficial lesions.

GCD1 deposits become more severe after radial keratotomy (RK). They are also more severe near the central portion of RK incisions, compared to the peripheral portion of RK incisions (Fig. 6.3) (Feizi et al. 2007). These findings suggest that deposition of abnormal transforming growth factor beta-induced protein (TGFBIp) is facilitated by the lack of vascularization in the central cornea.

Epidemiology and Inheritance

GCD1 is an autosomal dominant disorder associated with the p. (Arg555Trp) mutation in the transforming growth factor beta-induced (*TGFBI*) gene on chromosome 5q31 (Munier et al. 1997; Korvatska et al. 1998). According to published studies, p.Arg124His is the most frequently observed mutation in the Asian population. The p.Arg124His mutation has been found to be the most common, accounting for up to 72% of Japanese patients with corneal dystrophies (Mashima et al. 2000; Fujiki et al. 2001). The second most frequently observed mutation is probably GCD1 or lattice corneal dystrophy (LCD) type 1 (Mashima et al. 2000; Fujiki et al. 2001). In China, GCD1 is reportedly the most prevalent *TGFBI* mutation followed by LCD1 and GCD2 (Yang et al. 2010). Although a positive family history is nearly always obtained from individuals with *TGFBI* dystrophies, spontaneous mutations in the *TGFBI* gene have been reported in association with several of the *TGFBI* dystrophies, including granular corneal dystrophy (Hilton et al. 2007).

Histopathology

Light microscopy demonstrates eosinophilic rod or trapezoidal-shaped hyaline deposits beneath the epithelium and in the stroma. These deposits stain bright red with Masson's trichrome and stain weakly with periodic acid–Schiff (PAS) (Fig. 6.4) (Jones and Zimmerman 1961). The peripheral portions of the deposits also may stain with Congo red (Garner 1969; Iwamoto et al. 1975). Masson trichrome-stained keratohyalin deposits have been found at various stromal layers, and partial destruction of Bowman layer has also been noted. Immunohistochemical staining with antibodies to the TGFBIp (kerato-epithelin) demonstrates that the stromal deposits that characterize GCD1 consist of the mutated TGFBIp (Klintworth et al. 1998; Streeten et al. 1999).

Electron microscopy reveals rod-shaped or trapezoidal extracellular structures 100–500 μm wide, which may display a homogeneous, filamentous, or moth-eaten pattern in their inner structure (Fig. 6.5) (Sornson 1965; Akiya and Brown 1970; Iwamoto et al. 1975; Waring et al. 1978; Wittebol-Post et al. 1987). Although it is classified as a stromal dystrophy, studies suggest that its granular opacities may actually originate from the corneal epithelium and migrate to the corneal stroma (Waring et al. 1978; Johnson et al. 1981; Akhtar et al. 1999; Chiambaretta et al. 2004; Yang et al. 2010). Given the presence of these characteristic rod-shaped structures in both epithelial cells and stromal keratocytes, and because epithelial findings are prominent in recurrent granular dystrophy, it has been suggested that GCD1 arises in the epithelium (Garner 1969; Akiya and Brown 1970; Witschel and Sundmacher 1979; Johnson et al. 1981;

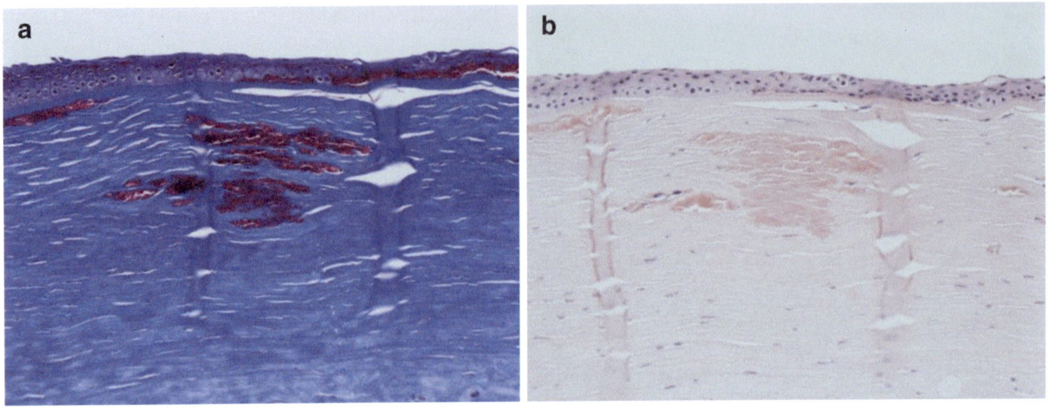

Fig. 6.4 GCD1 deposits beneath the epithelium and in the anterior stroma. These deposits stain bright red with Masson trichrome (**a**) and weakly with periodic acid–Schiff (**b**)

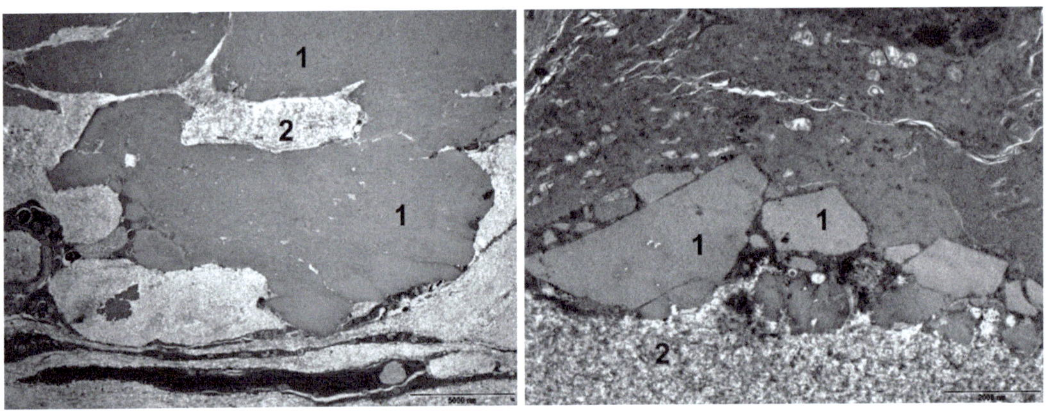

Fig. 6.5 Transmission electron micrograph demonstrates dense amorphous deposits and microfibrillar protein. 1: dense amorphous deposits, 2: microfibrillar protein

Wittebol-Post et al. 1987; Lyons et al. 1994). However, isolated recurrence of GCD1 deposits in the stroma posterior to the graft–host junction after anterior lamellar surgical interventions argues that the opacities originate in keratocytes (Auw-Haedrich et al. 1996; Pantanelli et al. 2014).

Surrounding these lesions may be 8–10 nm tubular microfibrils that usually lack the typical orientation of amyloid, despite Congo red staining (Fig. 6.5) (Rodrigues et al. 1983). Stromal keratocytes may be normal in appearance or may be in various stages of degeneration, with dilation of the endoplasmic reticulum and Golgi apparatus as well as vacuolation of the cytoplasm (Sornson 1965).

Management

Recurrent epithelial erosions should be managed routinely with therapeutic contact lenses and artificial tears. If vision is markedly reduced, surgical management may be considered, which varies depending on the depth and extent of the stromal lesions. If the opacities are superficial, superficial keratectomy may be performed (Lempert et al. 1978; Moller and Ehlers 1985; Lyons et al. 1994). Excimer laser phototherapeutic keratectomy (PTK) is the procedure of choice for the management of visually significant GCD1 deposits located in the anterior 100 microns of the corneal stroma (Stark et al. 1992; Hahn et al. 1993). PTK has also been performed in cases of superficial

GCD1 recurrence (Stewart et al. 2002); however, only short-term visual improvement can be achieved, as GCD1 recurrence after PTK is common (John et al. 1994).

In the setting of visually significant stromal deposits that are located more than 100 microns beneath the corneal surface, anterior lamellar keratoplasty (ALK) or penetrating keratoplasty (PKP) may be performed, although these procedures are not commonly required before the fifth decade of life. Granular dystrophy can recur in grafts as early as 1 year after surgery, and the recurrence-free interval seems to be independent of size and type of transplant performed (Herman and Hughes 1973; Brownstein et al. 1974; Rodrigues and McGavic 1975; Stuart et al. 1975; Lyons et al. 1994). The recurrence rate after PKP is reported to be 46.4% in GCD1 at 10 years (Ellies et al. 2002). Unlike previous studies, PKP provided a substantially longer significant recurrence-free interval (median time 13.7 years) than ALK (3.7 years), deep anterior lamellar keratoplasty (DALK, 3.2 years), or phototherapeutic keratectomy (PTK, 2.7 years). However, PTK is an inherently less invasive procedure that provides faster visual rehabilitation, when compared with either PKP or DALK (Lewis et al. 2017). More superficial lesions may be removed by PTK with restoration of good vision (Lempert et al. 1978).

The morphology of recurrent opacities after keratoplasty is often different than that of opacities seen in primary disease. Recurrent lesions after keratoplasty are typically diffuse, subepithelial opacities originating in the periphery of the graft (Fig. 6.6). Recurrence also may appear like more typical granular lesions in the middle and posterior portions of the donor stroma after penetrating keratoplasty (Brownstein et al. 1974; Rodrigues and McGavic 1975). In most cases, recurrences in grafts tend to be initially superficial and situated in the epithelial or subepithelial corneal layers (Witschel and Sundmacher 1979). The recurrence of the dystrophy first appears centrally and superficially, occasionally producing a vortex pattern. These results suggest epithelial involvement (Lyons et al. 1994).

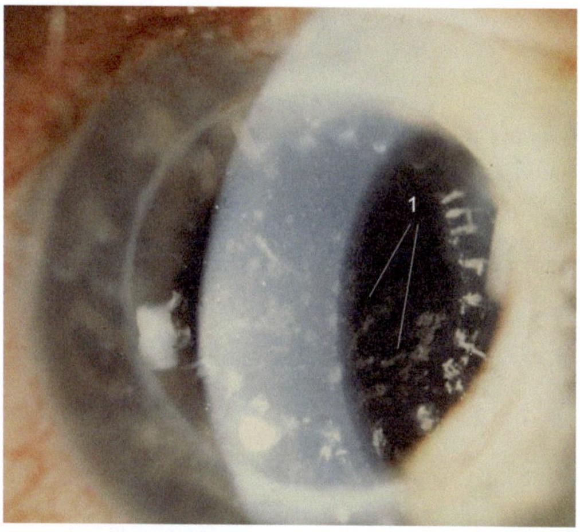

Fig. 6.6 Recurrence following corneal transplantation. 1: Diffuse subepithelial lesions of recurrence of branch shape which were denser in the periphery compared with center

References

Akhtar S, Meek KM, Ridgway AE, Bonshek RE, Bron AJ. Deposits and proteoglycan changes in primary and recurrent granular dystrophy of the cornea. Arch Ophthalmol. 1999;117(3):310–21. https://doi.org/10.1001/archopht.117.3.310.

Akiya S, Brown SI. Granular dystrophy of the cornea. Characteristic electron microscopic lesion. Arch Ophthalmol. 1970;84(2):179–92. https://doi.org/10.1001/archopht.1970.00990040181011.

Auw-Haedrich C, Loeffler KU, Sundmacher R, Witschel H. Characteristic distribution of deposits in recurrent granular corneal dystrophy. Ger J Ophthalmol. 1996;5(3):132–6.

Brownstein S, Fine BS, Sherman ME, Zimmerman LE. Granular dystrophy of the cornea. Light and electron microscopic confirmation of recurrence in a graft. Am J Ophthalmol. 1974;77(5):701–10.

Chiambaretta F, Pilon F, Deriot JB, Gerard M, Couleangon ML, Schorderet DF, Kemeny JL, Dastugue B, Creveaux I, Rigal D. Early recurrence of Groenouw type I corneal dystrophy after phototherapeutic keratectomy. Molecular biology study suggests epithelial genesis. J Fr Ophtalmol. 2004;27(5):449–56. https://doi.org/10.1016/s0181-5512(04)96164-5.

Ellies P, Renard G, Valleix S, Boelle PY, Dighiero P. Clinical outcome of eight BIGH3-linked corneal dystrophies. Ophthalmology. 2002;109(4):793–7. https://doi.org/10.1016/s0161-6420(01)01025-9.

Feizi S, Pakravan M, Baradaran-Rafiee AR, Yazdani S. Granular corneal dystrophy manifesting after radial keratotomy. Cornea. 2007;26(10):1267–9. https://doi.org/10.1097/ICO.0b013e3181450fa8.

Forsius H, Eriksson AW, Karna J, Tarkkanen A, Aurekoski H, Frants RR, Damsten M. Granular corneal dystrophy with late manifestation. Acta Ophthalmol (Copenh). 1983;61(4):514–28. https://doi.org/10.1111/j.1755-3768.1983.tb04341.x.

Fujiki K, Nakayasu K, Kanai A. Corneal dystrophies in Japan. J Hum Genet. 2001;46(8):431–5. https://doi.org/10.1007/s100380170041.

Garner A. Histochemistry of corneal granular dystrophy. Br J Ophthalmol. 1969;53(12):799–807. https://doi.org/10.1136/bjo.53.12.799.

Haddad R, Font RL, Fine BS. Unusual superficial variant of granular dystrophy of the cornea. Am J Ophthalmol. 1977;83(2):213–8. https://doi.org/10.1016/0002-9394(77)90619-5.

Hahn TW, Sah WJ, Kim JH. Phototherapeutic keratectomy in nine eyes with superficial corneal diseases. Refract Corneal Surg. 1993;9(2 Suppl):S115–8.

Herman SJ, Hughes WF. Recurrence of hereditary corneal dystrophy following keratoplasty. Am J Ophthalmol. 1973;75(4):689–94. https://doi.org/10.1016/0002-9394(73)90822-2.

Hilton EN, Black GC, Manson FD, Schorderet DF, Munier FL. De novo mutation in the BIGH3/TGFB1 gene causing granular corneal dystrophy. Br J Ophthalmol. 2007;91(8):1083–4. https://doi.org/10.1136/bjo.2006.103283.

Iwamoto T, Stuart JC, Srinivasan BD, Mund ML, Farris RL, Donn A, DeVoe AG. Ultrastructural variation in granular dystrophy of the cornea. Albrecht Von Graefes Arch Klin Exp Ophthalmol. 1975;194(1):1–9. https://doi.org/10.1007/BF00408270.

John ME, Martines E, Cvintal T, Mellor Filho A, Soter F, Barbosa de Sousa MC, Boleyn KL, Ballew C. Photorefractive keratectomy following penetrating keratoplasty. J Refract Corneal Surg. 1994;10(2 Suppl):S206–10.

Johnson BL, Brown SI, Zaidman GW. A light and electron microscopic study of recurrent granular dystrophy of the cornea. Am J Ophthalmol. 1981;92(1):49–58. https://doi.org/10.1016/s0002-9394(14)75907-0.

Jones ST, Zimmerman LE. Histopathologic differentiation of granular, macular and lattice dystrophies of the cornea. Am J Ophthalmol. 1961;51:394–410. https://doi.org/10.1016/0002-9394(61)92085-2.

Klintworth GK, Valnickova Z, Enghild JJ. Accumulation of beta ig-h3 gene product in corneas with granular dystrophy. Am J Pathol. 1998;152(3):743–8.

Korvatska E, Munier FL, Djemai A, Wang MX, Frueh B, Chiou AG, Uffer S, Ballestrazzi E, Braunstein RE, Forster RK, Culbertson WW, Boman H, Zografos L, Schorderet DF. Mutation hot spots in 5q31-linked corneal dystrophies. Am J Hum Genet. 1998;62(2):320–4. https://doi.org/10.1086/301720.

Lempert SL, Jenkins MS, Johnson BL, Brown SI. A simple technique for removal of recurring granular dystrophy in corneal grafts. Am J Ophthalmol. 1978;86(1):89–91. https://doi.org/10.1016/0002-9394(78)90020-x.

Lewis DR, Price MO, Feng MT, Price FW Jr. Recurrence of granular corneal dystrophy type 1 after phototherapeutic keratectomy, lamellar keratoplasty, and penetrating keratoplasty in a single population. Cornea. 2017;36(10):1227–32. https://doi.org/10.1097/ICO.0000000000001303.

Lyons CJ, McCartney AC, Kirkness CM, Ficker LA, Steele AD, Rice NS. Granular corneal dystrophy. Visual results and pattern of recurrence after lamellar or penetrating keratoplasty. Ophthalmology. 1994;101(11):1812–7. https://doi.org/10.1016/s0161-6420(94)31096-7.

Mashima Y, Yamamoto S, Inoue Y, Yamada M, Konishi M, Watanabe H, Maeda N, Shimomura Y, Kinoshita S. Association of autosomal dominantly inherited corneal dystrophies with BIGH3 gene mutations in Japan. Am J Ophthalmol. 2000;130(4):516–7. https://doi.org/10.1016/s0002-9394(00)00571-7.

Moller HU. Granular corneal dystrophy Groenouw type I. Clinical and genetic aspects. Acta Ophthalmol Suppl (1985). 1991;198:1–40.

Moller HU, Ehlers N. Early treatment of granular dystrophy (Groenouw type I). Acta Ophthalmol (Copenh). 1985;63(5):597–600. https://doi.org/10.1111/j.1755-3768.1985.tb05252.x.

Munier FL, Korvatska E, Djemai A, Le Paslier D, Zografos L, Pescia G, Schorderet DF. Kerato-epithelin mutations in four 5q31-linked corneal dystrophies. Nat Genet. 1997;15(3):247–51. https://doi.org/10.1038/ng0397-247.

Okada M, Yamamoto S, Watanabe H, Inoue Y, Tsujikawa M, Maeda N, Shimomura Y, Nishida K, Kinoshita S, Tano Y. Granular corneal dystrophy with homozygous mutations in the kerato-epithelin gene. Am J Ophthalmol. 1998;126(2):169–76. https://doi.org/10.1016/s0002-9394(98)00075-0.

Pantanelli SM, Herzlich A, Yeaney G, Ching ST. Recurrence of granular corneal dystrophy type I deposits within host stroma after non-descemet baring anterior lamellar keratoplasty. Cornea. 2014;33(12):1348–51. https://doi.org/10.1097/ICO.0000000000000267.

Rodrigues MM, McGavic JS. Recurrent corneal granular dystrophy: a clinicopathologic study. Trans Am Ophthalmol Soc. 1975;73:306–16.

Rodrigues MM, Streeten BW, Krachmer JH, Laibson PR, Salem N Jr, Passonneau J, Chock S. Microfibrillar protein and phospholipid in granular corneal dystrophy. Arch Ophthalmol. 1983;101(5):802–10. https://doi.org/10.1001/archopht.1983.01040010802023.

Rodrigues MM, Gaster RN, Pratt MV. Unusual superficial confluent form of granular corneal dystrophy. Ophthalmology. 1984;91(1 Suppl):1507–11.

Sekimoto M, Hayasaka S, Furuse N, Setogawa T. Concomitant granular dystrophy of the cornea and cone dystrophy. Ophthalmologica. 1988;197(2):57–9. https://doi.org/10.1159/000309920.

Sornson ET. Granular dystrophy of the cornea: an electron microscopic study. Am J Ophthalmol. 1965;59:1001–7.

Stark WJ, Chamon W, Kamp MT, Enger CL, Rencs EV, Gottsh JD. Clinical follow-up of 193-nm ArF excimer laser photokeratectomy. Ophthalmology. 1992;99(5):805–12. https://doi.org/10.1016/s0161-6420(92)31896-2.

Stewart OG, Pararajasegaram P, Cazabon J, Morrell AJ. Visual and symptomatic outcome of excimer phototherapeutic keratectomy (PTK) for corneal dystrophies. Eye (Lond). 2002;16(2):126–31. https://doi.org/10.1038/sj.eye.6700049.

Streeten BW, Qi Y, Klintworth GK, Eagle RC Jr, Strauss JA, Bennett K. Immunolocalization of beta ig-h3 protein in 5q31-linked corneal dystrophies and normal corneas. Arch Ophthalmol. 1999;117(1):67–75. https://doi.org/10.1001/archopht.117.1.67.

Stuart JC, Mund ML, Iwamoto T, Troutman RC, White H, DeVoe AG. Recurrent granular corneal dystrophy. Am J Ophthalmol. 1975;79(1):18–24. https://doi.org/10.1016/0002-9394(75)90451-1.

Waring GO 3rd, Rodrigues MM, Laibson PR. Corneal dystrophies. I. Dystrophies of the epithelium, Bowman's layer and stroma. Surv Ophthalmol. 1978;23(2):71–122. https://doi.org/10.1016/0039-6257(78)90090-5.

Weidle EG, Lisch W. Various forms of opacities of granular corneal dystrophy. Klin Monbl Augenheilkd. 1984;185(3):167–73. https://doi.org/10.1055/s-2008-1054592.

Weiss JS, Moller HU, Aldave AJ, Seitz B, Bredrup C, Kivela T, Munier FL, Rapuano CJ, Nischal KK, Kim EK, Sutphin J, Busin M, Labbe A, Kenyon KR, Kinoshita S, Lisch W. IC3D classification of corneal dystrophies—edition 2. Cornea. 2015;34(2):117–59. https://doi.org/10.1097/ICO.0000000000000307.

Witschel H, Sundmacher R. Bilateral recurrence of granular corneal dystrophy in the grafts. A clinico-pathologic study. Albrecht Von Graefes Arch Klin Exp Ophthalmol. 1979;209(3):179–88. https://doi.org/10.1007/BF00414610.

Wittebol-Post D, van der Want JJ, van Bijsterveld OP. Granular dystrophy of the cornea (Groenouw's type I). Is the keratocyte the primary source after all? Ophthalmologica. 1987;195(4):169–77. https://doi.org/10.1159/000309808.

Yang J, Han X, Huang D, Yu L, Zhu Y, Tong Y, Zhu B, Li C, Weng M, Ma X. Analysis of TGFBI gene mutations in Chinese patients with corneal dystrophies and review of the literature. Mol Vis. 2010;16:1186–93.

7 Granular Corneal Dystrophy Type 2 (GCD2)

Ikhyun Jun and Eung Kweon Kim

Granular corneal dystrophy type 2 (GCD2) is an autosomal dominant dystrophy with very high penetrance, which is characterized by granular and linear deposits in the deep stroma in early stages and superficial diffuse haze in advanced stages (Han et al. 2010; Weiss et al. 2015). In 1938, Bücklers described a large family with the phenotype of GCD2, which has long been considered a mild variant of GCD1 (Weiss et al. 2015). Weidle reported and classified GCD patients according to subtle clinical differences in the 1980s. Folberg et al. demonstrated the histopathology of clinically atypical GCD with granular deposits in the anterior third of the stroma, and lattice-like amyloid deposits in the deeper stroma, which is now known as GCD2 (Folberg et al. 1988). In 1992, this dystrophy was separated from GCD and named Avellino corneal dystrophy, because the pedigree that was reported came from the Avellino district of Italy (Holland et al. 1992). This entity had been also described as combined granular-lattice corneal dystrophy before when the granular and linear lesions were observed in the same cornea together (Holland et al. 1992; Rosenwasser et al. 1993; Akiya et al. 1999). GCD2 is found worldwide now and is reported with greater frequently in East Asia (Korea and Japan) than in Italy (Lisch and Weiss 2019). The term GCD2 is consistent with the recommendations of the International Committee for Classification of Corneal Dystrophies (IC3D), published in 2008 (Weiss et al. 2008).

Genetics and Epidemiology

GCD2 is an autosomal dominant disorder that is caused by the p.Arg124His mutation of the transforming growth factor beta-induced (*TGFBI*) gene on chromosome 5q31 (Munier et al. 1997; Korvatska et al. 1998). The phenotype of heterozygotes is milder than that of homozygotes. Other mutations of the Arg 124 residue cause Reis-Bücklers corneal dystrophy (RBCD; p.Arg124Leu), LCD1 (p.Arg124Cys) (Han et al. 2016; Stewart et al. 1999), and GCD1 (p.Arg124Ser) (Bouyacoub et al. 2019).

I. Jun (✉)
Department of Ophthalmology, Yonsei University College of Medicine, Seoul, Korea (Republic of)

The Corneal Dystrophy Research Institute, Yonsei University College of Medicine, Seoul, Korea (Republic of)

Institute of Vision Research, Yonsei University College of Medicine, Seoul, Korea (Republic of)
e-mail: hadesdual@yuhs.ac

E. K. Kim
Department of Ophthalmology, Yonsei University College of Medicine, Seoul, Korea (Republic of)

The Corneal Dystrophy Research Institute, Yonsei University College of Medicine, Seoul, Korea (Republic of)

Saevit Eye Hospital, Goyang, Korea (Republic of)
e-mail: eungkkim@yuhs.ac

E. K. Kim, R. D. Stulting (eds.), *TGFBI-related Corneal Dystrophies*,
https://doi.org/10.1007/978-981-96-0131-8_7

TGFBI mutations other than that of p. Arg124His may alter the severity or, rarely, the phenotype of GCD2. These include p.Leu509Pro (Gruenauer-Kloevekorn et al. 2009), p.Leu550Pro (Lakshminarayanan et al. 2011; Zenteno et al. 2009), and p.Met619Lys (Aldave et al. 2008).

The prevalence of GCD2 varies by country and race. Reports from Korea and Japan estimate that GCD2 accounts for 72–91% of the *TGFBI*-related corneal dystrophies (Mashima et al. 2000; Yamamoto et al. 2000; Kim et al. 2001; Cho et al. 2012), and they account for about 67% of them in China (Yu et al. 2003). In contrast, Afshari et al. reported that 5 of 14 cases (36%) of *TGFBI*-related corneal dystrophies were GCD2 in the United States (Afshari et al. 2001). In a Polish population, there were only 2 cases of GCD2 among 60 *TGFBI*-related corneal dystrophy patients (Nowinska et al. 2011). Lee et al. estimated the prevalence of GCD2 among the South Korean population to be 11.5 or more affected persons per 10,000 individuals (0.115%) (Lee et al. 2010). Musch et al. estimated the prevalence of GCD (not GCD2) to be 0.002% in the United States by reviewing medical records from a national managed-care network. Since this study did not provide an estimate of GCD2 specifically, the prevalence of GCD2 is undoubtedly lower. Notably, the prevalence of GCD in blacks was relatively low in the study (Musch et al. 2011).

Clinical Manifestations

Heterozygous patients can be diagnosed as early as age 8, but most patients are diagnosed as teenagers or young adults. Usually, heterozygous GCD2 is expressed as mild disease with minimal visual disturbance that is insignificant until late in life. Homozygotic GCD2 may be diagnosed as early as age 3 by the appearance of white corneal deposits that can be seen with the naked eye. Homozygotic GCD2 is visually significant from early childhood (Moon et al. 2007).

GCD2 is characterized by (1) well-circumscribed, gray-white granular deposits in the anterior stroma at an early stage, (2) lattice-like linear deposits in the mid to posterior stroma with progression, and (3) diffuse anterior stromal haze in later, more severe cases (Fig. 7.1) (Holland et al. 1992). The initial clinical manifestation of GCD2 is small whitish deposits in the superficial stroma (Fig. 7.1a), which enlarge to become granular opacities (Fig. 7.1b). As the opacities enlarge, round opacities become discoid in shape or ring-shaped when the central portion of the deposit fades spontaneously (Fig. 7.1c). In addition to the round opacities, most patients have stellate or spike-like opacities in the anterior and mid corneal stroma, which are partially translucent in retroillumination. As the disease progresses, the granular opacities become larger and more prominent, often becoming stellate, spiked, or linear opacities (Fig. 7.1d). Linear lesions usually do not appear before granular lesions, and they also become more apparent over time. Linear lesions are found deeper in the stroma than granular lesions (Fig. 7.1e). They can be distinguished from the lattice lines observed in classical LCD. The linear lesions of GCD2 are whiter and more opaque than those of LCD and less refractile than those of LCD. The granular lesions of GCD2 are composed of hyaline, and the linear lesions also contain amyloid. Detailed descriptions and discussions can be found in Chap. 9, "The Role of Extracellular Matrix in *TGFBI*-related Corneal Dystrophy Development" by Jan J. Enghild.

The linear lesions of GCD2 branch but rarely intersect, while lattice lines characteristically cross each other. In older patients with more advanced GCD, diffuse stromal haze appears in the anterior stroma beneath Bowman's layer, becomes denser, and is the major cause of reduced vision in older patients (Fig. 7.1f). Patients with GCD2 usually have fewer granular deposits than patients with GCD1.

When anterior granular GCD2 deposits invade Bowman's layer, the superficial granules are spontaneously exposed and extruded, resulting in corneal epithelial erosions, which cause pain, foreign body sensation, discomfort, and photophobia (Han et al. 2013). Recurrent corneal erosions occur more frequently in patients with GCD1 compared to those with GCD2 (Lisch and Weiss 2019). As portions of the granular deposits are extruded, ring-shaped opacities with punched-out centers are created; and as the opacity grows

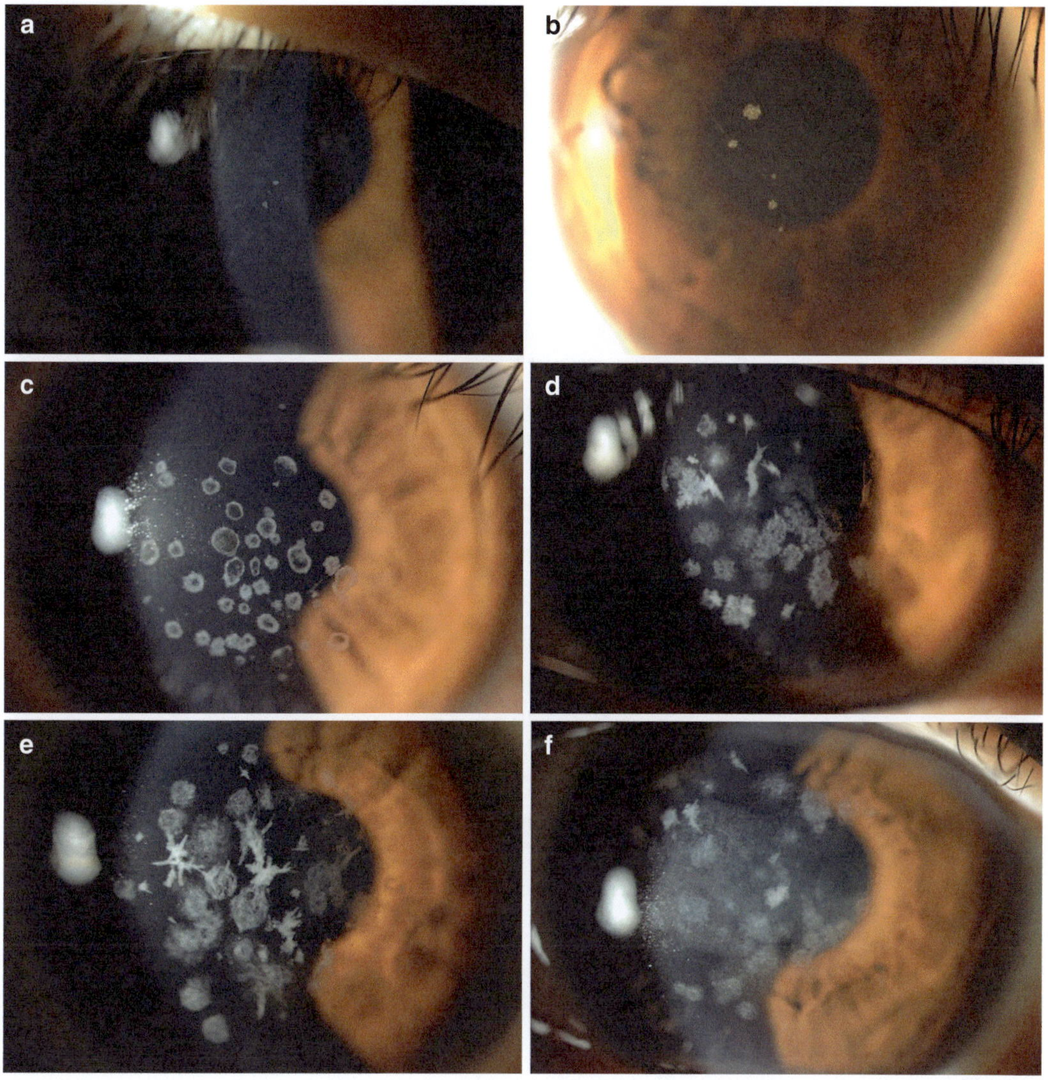

Fig. 7.1 Clinical manifestation of granular corneal dystrophy type 2 (GCD2). (**a**) A 10-year-old heterozygous GCD2 patient with a few white punctate stromal opacities. (**b**) A 23-year-old heterozygous GCD2 patient with some small round opacities in the anterior corneal stroma. (**c**) A 21-year-old heterozygous GCD2 patient with multiple round opacities that have punched-out centers, creating ring-shaped deposits. (**d**) A 43-year old heterozygous GCD2 patient with larger granular deposits and linear opacities. (**e**) A 64-year-old heterozygous GCD2 patient with lattice lines and large granular deposits. (**f**) A 70-year-old heterozygous GCD2 patient with advanced disease showing prominent diffuse stromal haze

again, rings become thicker; and the punched-out center again becomes opaque (Han et al. 2013).

The homozygous form of GCD2 shows numerous small spots in the cornea in early childhood. These opacities increase in size with age and tend to coalescence earlier than the opacities in heterozygotes (Fig. 7.2). The homozygote has two distinct phenotypes: (1) gray-white spot-like with confluent opacities (type I opacity, Fig. 7.2) (Moon et al. 2007; Mashima et al. 1998; Okada et al. 1998) and (2) reticular, grayish, confluent opacities with translucent intervening spaces in the anterior stroma (type 2 opacity, Fig. 7.3) (Tsujikawa et al. 2007; Watanabe et al. 2001). Because patients with the type 2 phenotype have translucent spaces between opacities, best corrected visual acuity tends to be better than it is in patients with the type 1 phenotype (Fig. 7.3). In homozygous GCD2,

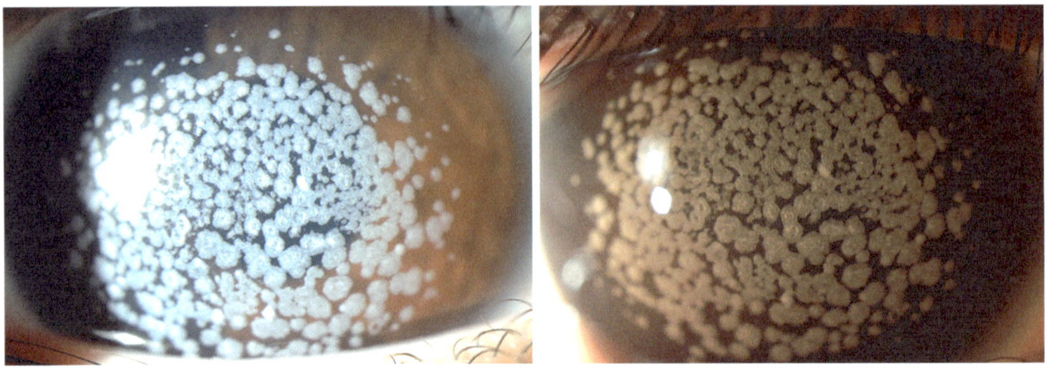

Fig. 7.2 Slit lamp photograph of a 6-year-old homozygous GCD2 patient with type 1 opacities using direct illumination and sclerotic scatter

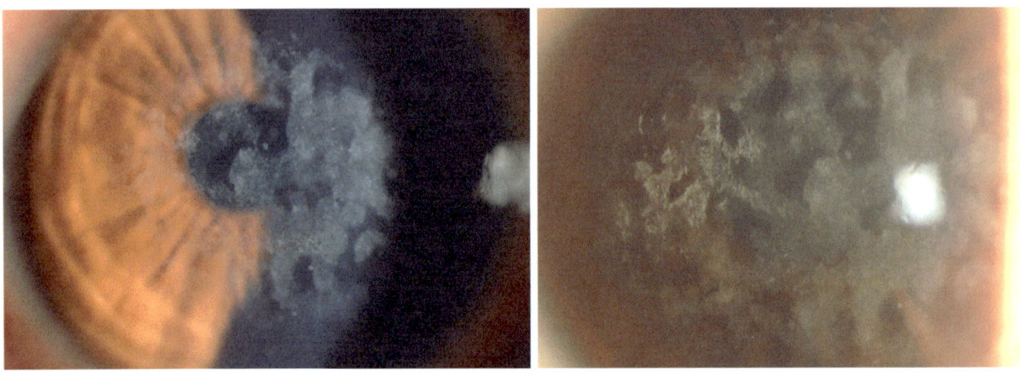

Fig. 7.3 Slit lamp photograph if a 20-year-old homozygous GCD2 patient with type 2 opacities using direct illumination and sclerotic scatter

lattice-like opacities are not found, and vision is often so impaired in early childhood that surgical intervention is required. Recurrence is frequent and rapid in homozygous patients undergoing phototherapeutic keratectomy (PTK) or corneal transplantation (Moon et al. 2007).

The severity of heterozygous GCD2 (p.Arg124His mutation) varies from a few, subtle white dots to multiple dense opacities throughout the stroma even at a relatively young age (Rosenwasser et al. 1993; Han et al. 2012; Abazi et al. 2013). Double mutations in *TGFBI* showing more severe phenotypes than those seen with a single mutation have been reported (Zenteno et al. 2009; Dighiero et al. 2000; Sakimoto et al. 2003; Klintworth et al. 2004; Aldave et al. 2004; Frising et al. 2006; Yamada et al. 2009; Zhong et al. 2010; Niel-Butschi et al. 2011; Yam et al. 2012; Song et al. 2015; Chae et al. 2016; Ann et al. 2017). Double mutation of p.Arg124His and p.Asn544Ser of the *TGFBI* gene caused combined granular opacity and LCD-like opacities in affected patients (Yamada et al. 2009). Patients heterozygous for the p.Arg124His and p.Leu103Leu_fs*28 (c.307_308delCT) mutation showed multiple large whitish granular discoid deposits without lattice-like opacities which resemble those of homozygous p.Arg124His GCD2 patients (Yam et al. 2012). Patients with p.Arg124His and p. Ala179* mutations in *TGFBI* gene showed a similar dense, confluent granular pattern without lattice-like opacities (Song et al. 2015).

Exacerbation of GCD2 after uncomplicated laser-assisted in situ keratomileusis (LASIK) was first reported in 2002 (Wan et al. 2002). New, severe granular corneal opacities appeared in the interface between the flap and the stromal bed (Fig. 7.4). Since then, many reports have described the exacerbation of GCD2 after LASIK surgery (Jun et al. 2004; Roh et al. 2006a;

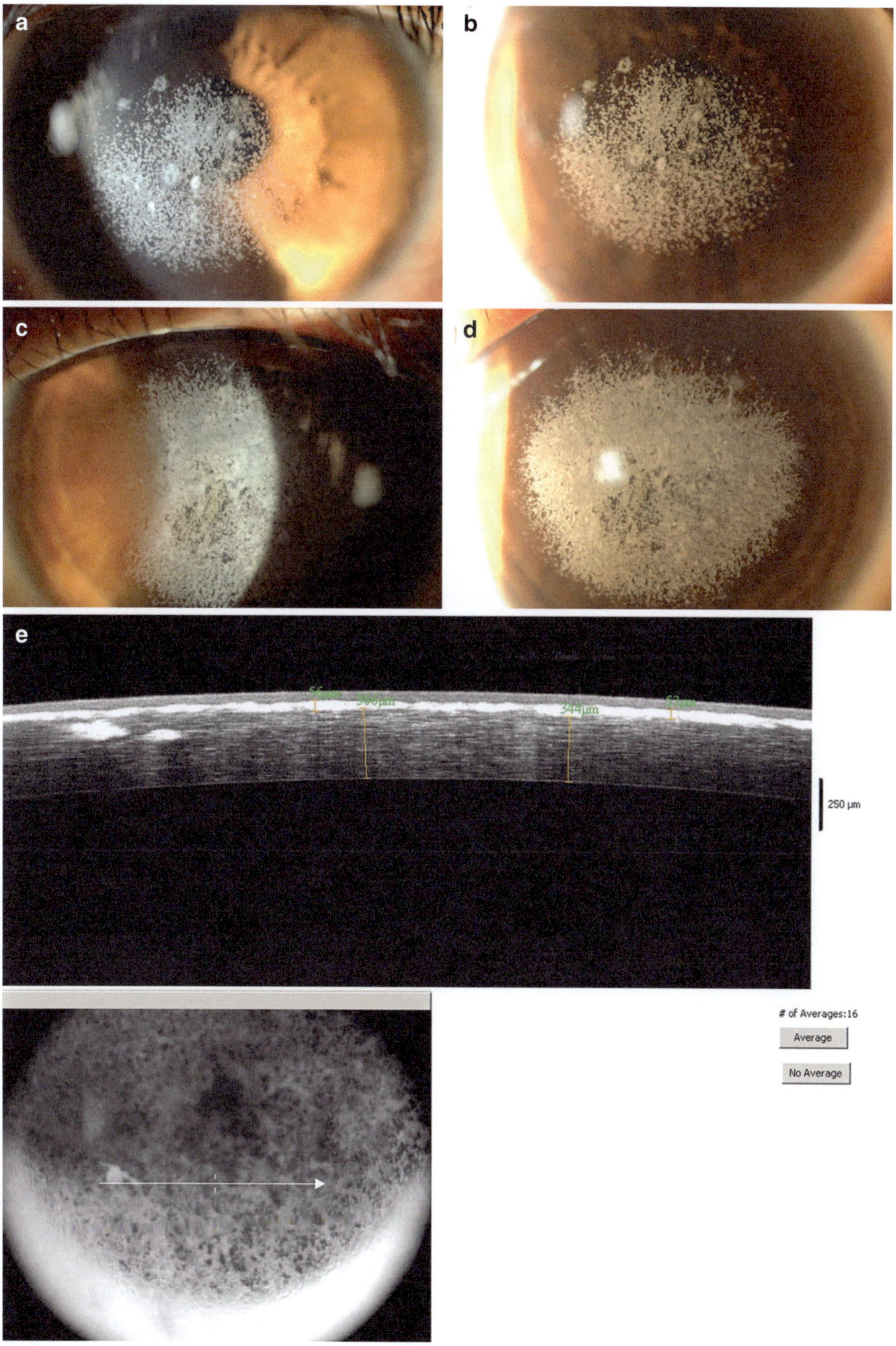

Fig. 7.4 Exacerbation of GCD2 after LASIK. (**a, b**) Shown with direct illumination and sclerotic scatter in a patient with GCD2 exacerbation 10 years after LASIK. (**c, d**) Slit lamp photograph of another patient with exacerbation of GCD2 13 years after LASIK using direct illumination and sclerotic scatter. (**e**) Anterior segment OCT illustrating dense white deposits in the flap interface

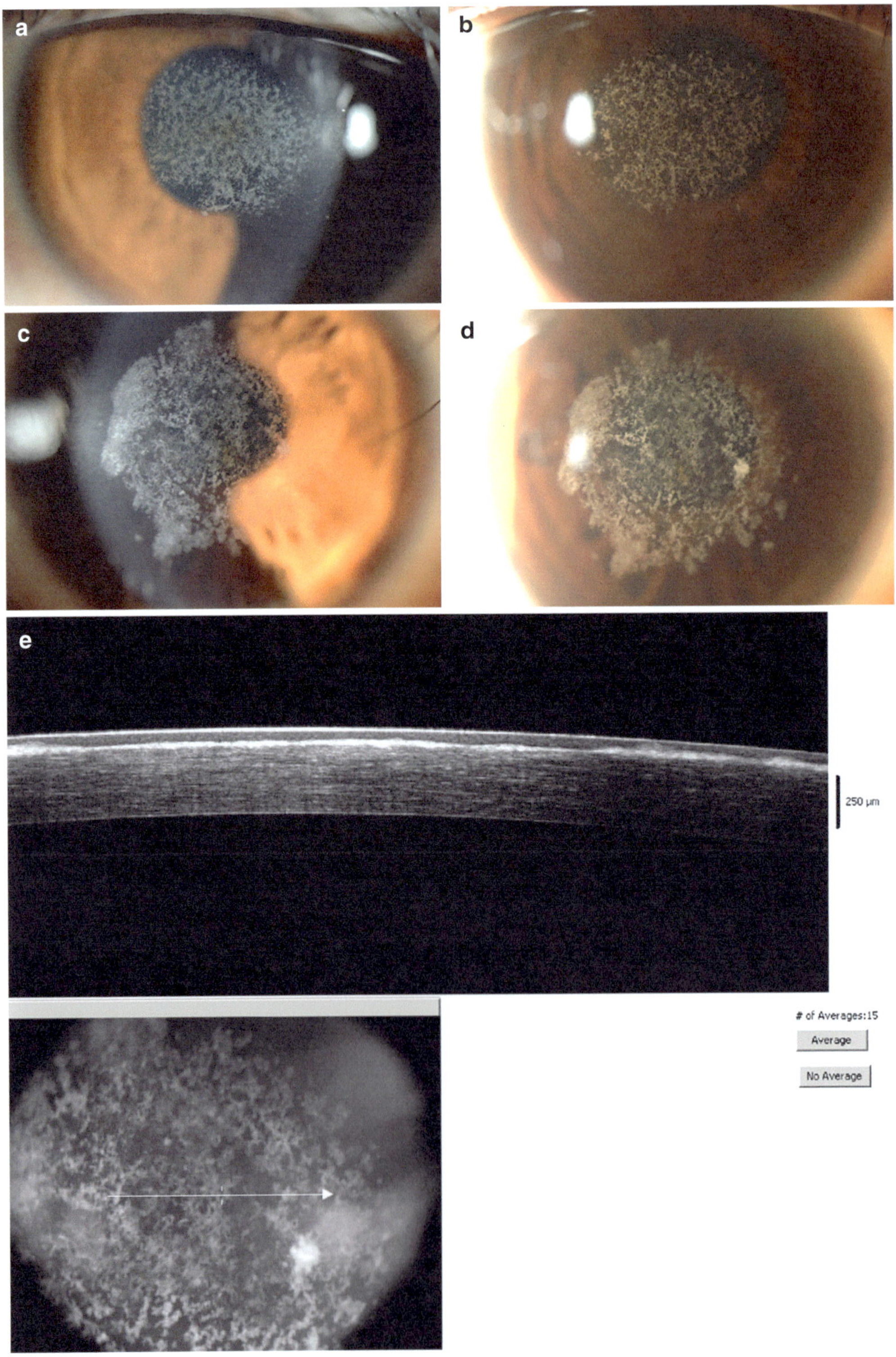

Fig. 7.5 Exacerbation of GCD2 after laser-assisted subepithelial keratectomy (LASEK). (**a, b**) Slit lamp photograph of a patient with GCD2 exacerbation 7 years after LASEK using direct illumination and sclerotic scatter. (**c, d**) Slit lamp photograph of another patient with GCD2 exacerbation 12 years after LASEK using direct illumination and sclerotic scatter. (**e**) Anterior segment OCT illustrating deposits primarily in the anterior stroma beneath the epithelium

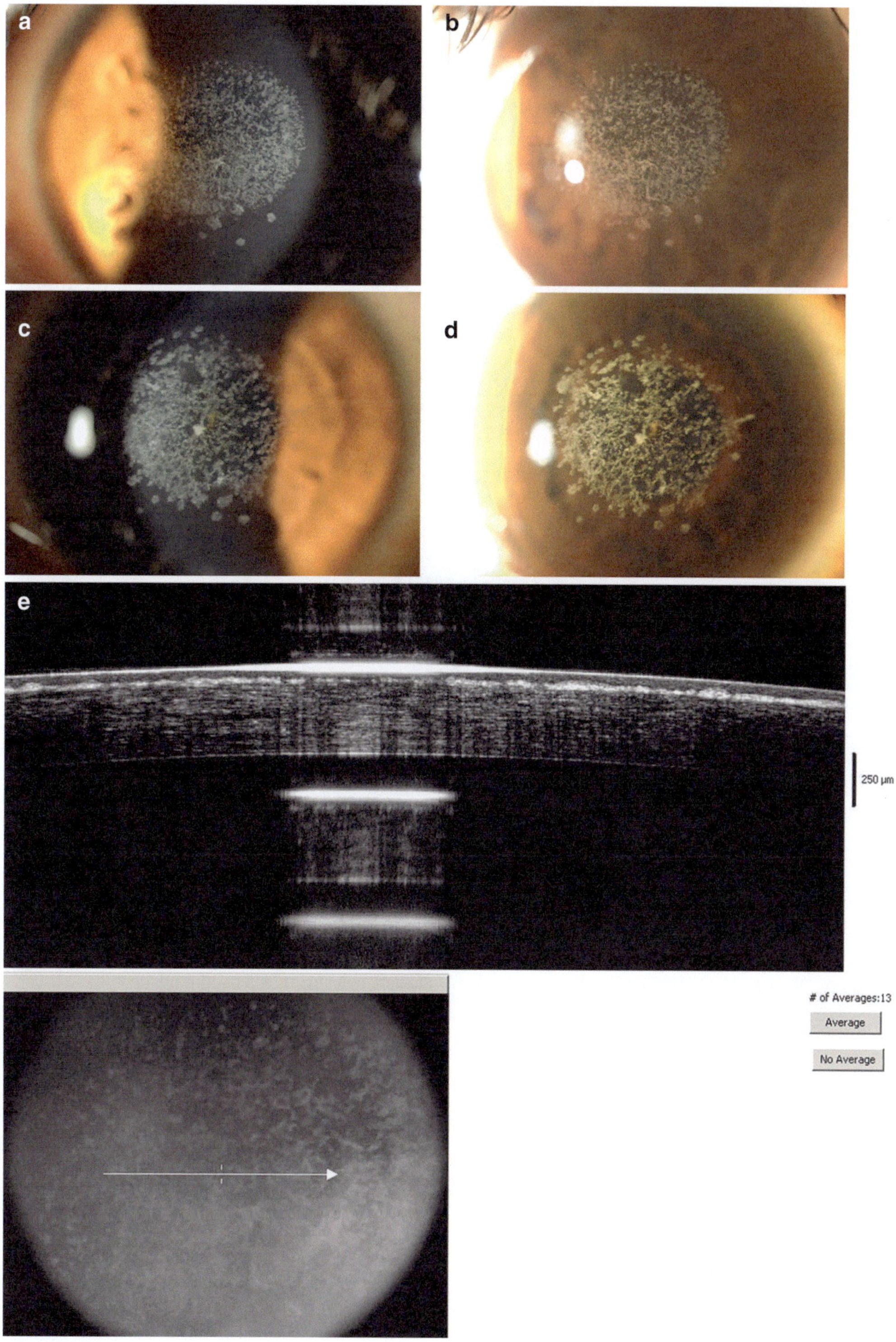

Fig. 7.6 Exacerbation of GCD2 after photorefractive keratectomy (PRK). (**a, b**) Slit lamp photograph of a patient with GCD2 exacerbation 8 years after PRK using direct illumination and sclerotic scatter. (**c, d**) Slit lamp photograph of another patient with GCD2 exacerbation 13 years after PRK using direct illumination and sclerotic scatter. (**e**) Anterior segment OCT showing deposits primarily in the anterior stroma beneath the epithelium

Banning et al. 2006; Aldave et al. 2007). Exacerbation of GCD2 has also been observed after PRK and LASEK (Figs. 7.5 and 7.6) (Kim et al. 2008a; Lee et al. 2008). In a study comparing LASIK and PRK (Kim et al. 2008a), it was reported that corneal opacities are more severe and visual acuity worse following LASIK than they are following PRK (Fig. 7.7). A recent report of GCD2 exacerbation after small incision lenticule extraction (SMILE) shows that traumatizing the central corneal stroma with minimal disruption of the epithelial or Bowman's layer is also contraindicated in patients with GCD2 (Kwak et al. 2021).

GCD typically recurs between 1 and 3 years after LASEK, which is thought to be later than after LASIK (Lee et al. 2008). After LASIK, multiple small, granular deposits composed of hyaline material appear in the interface between the flap and stromal bed. After LASEK and PRK, numerous fine deposits are found in the superficial stroma.

GCD deposits are typically found in the central cornea, with a clear area near the limbus. In GCD2 homozygotes, the width of the peripheral lucid interval decreases with age (Fig. 7.8). When a pterygium is present in GCD2 heterozygotes, the deposits are more distant from the lim-

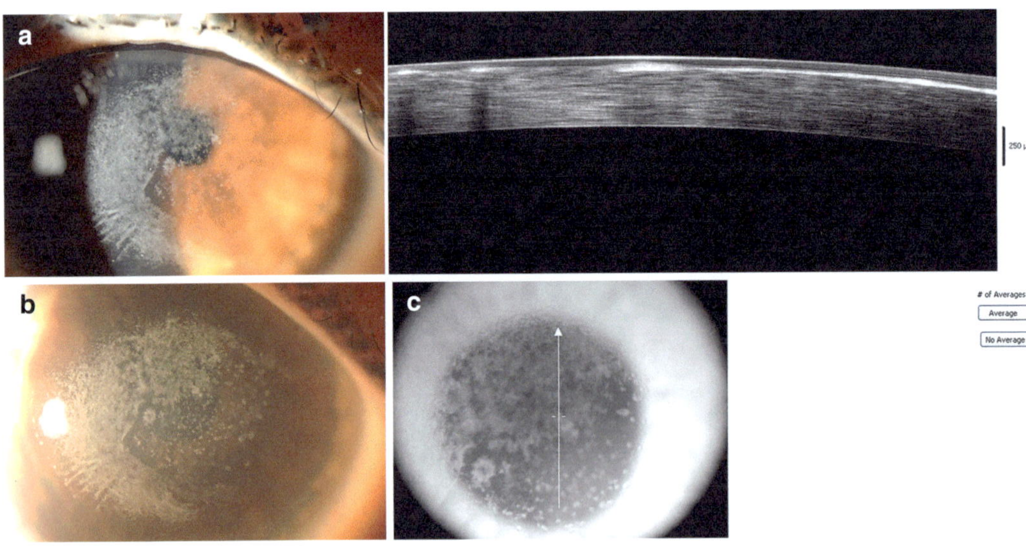

Fig. 7.7 A 39-year-old GCD2 patient who underwent LASIK complicated by an incomplete and irregular flap. Coarse opacities on the right side, and scant deposits on the left side using direct illumination (**a**) and sclerotic scatter (**b**). (**c**) Anterior segment OCT showing that the dense white deposits are in the flap interface

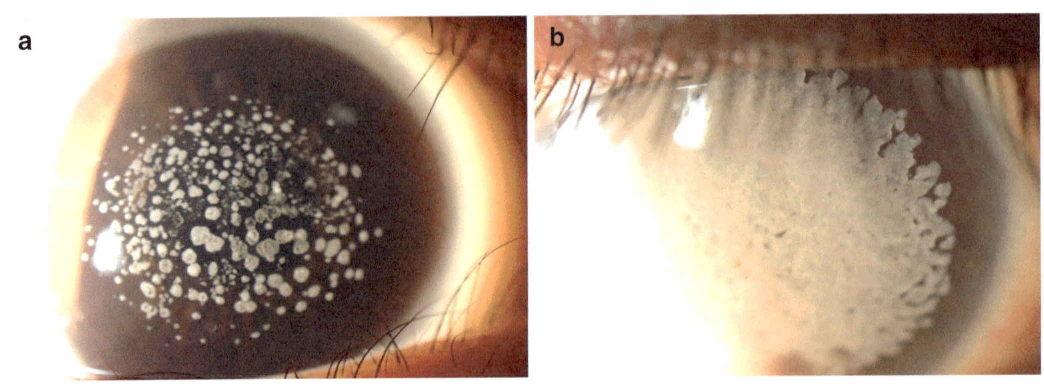

Fig. 7.8 Slit lamp photographs of homozygous GCD2 patients ((**a**) 3-year-old; (**b**) 46-year-old) showing lucid intervals between the opacities and the limbus. The width of the lucid interval decreases with age

bus (3.5–4.8 mm) near the pterygium than they are on the opposite side of the cornea away from the pterygium (2.8 ± 0.57 mm) (Fig. 7.9).

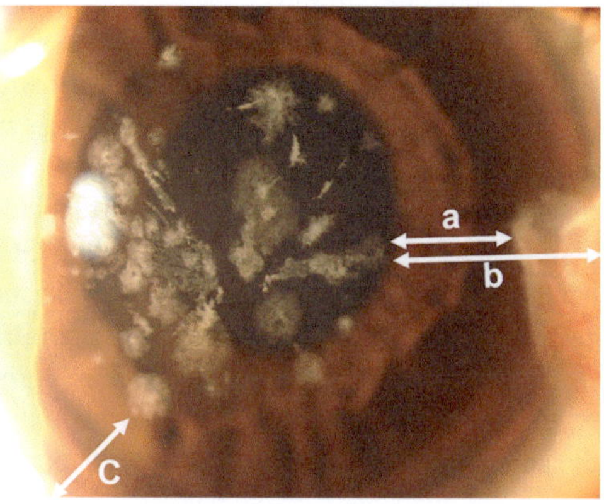

Fig. 7.9 Distance between opacities and the limbus in a GCD2 patient with pterygium. The distance between nasal opacities and the nasal pterygium (**a**) is similar to the distance between temporal opacities and the temporal limbus (**c**). The distance between nasal opacities and the nasal limbus (**b**) is greater than the distance between temporal opacities and the temporal limbus (**c**)

Indistinct deposits near the head of the pterygium suggest that the deposits are being resorbed (Lee et al. 2006). The peripheral cornea of a patient with GCD2 and phthisis with corneal neovascularization had no GCD deposits (Fig. 7.10).

LASIK surgery performed in patients with GCD2 frequently results in exacerbation of the disease, yet clear corneal incisions near the limbus performed during cataract surgery do not exacerbate GCD2 (Roh et al. 2006b). This observation supports the hypothesis that GCD deposits are cleared by some characteristic of the vascular supply. We also suspect the narrowing of the lucid interval in the aging homozygote might be related, at least in part, to the decreased ability of the GCD2 keratocyte to clear mutated transforming growth factor beta-induced protein (TGFBIp) because the aging process of the GCD2 keratocyte is accelerated, compared to that of the normal cell (Choi et al. 2012).

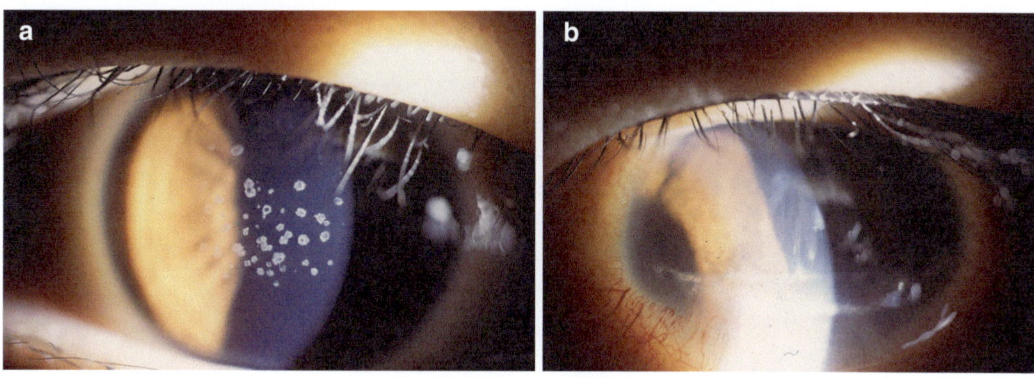

Fig. 7.10 Slit lamp photographs of a patient with GCD2 and phthisis and corneal neovascularization. Granular opacities are present in the right eye (**a**), but not in the left eye (**b**)

Histopathology

Light microscopy of corneas from patients with GCD2 demonstrates sharply demarcated eosinophilic deposits from the basal epithelium through the anterior stroma and multiple fusiform, pale, eosinophilic deposits in the mid—to deep stroma with hematoxylin-eosin stain (Folberg et al. 1988). These hyaline and amyloid deposits stain with Masson trichrome and/or Congo red (Fig. 7.11a, b). Typically, hyaline deposits are located in the superficial stroma, and amyloid deposits are located in deep stroma, but amyloid patches have also been observed in the Bowman's layer (Streeten et al. 1999). Amyloid deposits stain with Congo red. They also demonstrate birefringence and dichroism with cross-polarization (Fig. 7.11c).

Transmission electron microscopy of corneas from patients with GCD2 demonstrates rod-shaped electron-dense deposits in the anterior stroma, resembling those seen with GCD1 (Fig. 7.11d) (Weiss et al. 2015). However, they are not as heavily labeled for TGFBIp as they are in GCD1, and they show lucent spaces surrounded by well-labeled amorphous material (Streeten et al. 1999). Under higher magnification, the rod-shaped deposits are featureless, coalescing with the adjacent fine fibrillo-granular material. The amyloid deposits located in the deep stroma are strongly labeled with TGFBIp and infiltrated between collagen bundles similar to those seen in LCD1. Linear fibrils of the superficial stroma are less intensely labeled than stromal amyloid and mixed with fine TGFBIp-labeled materials. In normal keratocytes, it is difficult to distinguish intracellular organelles, while GCD2 heterozygous keratocytes are enlarged and intracellular organelles can be more easily identified.

Confocal microscopy of GCD2 corneas reveals both GCD1—and LCD-like patterns. In the anterior stroma, hyper-reflective breadcrumb-like round, well-demarcated deposits, or irregular snowflake-shaped deposits are observed similar to those seen in GCD1. Linear and branching highly reflective structures similar to those seen in LCD are seen in the deep stroma (Fig. 7.11e) (Weiss et al. 2015).

Severe recurrences of GCD2 after LASIK are primarily due to deposits in the flap interface, which stain with Masson trichrome, but not Congo red, suggesting that the main component

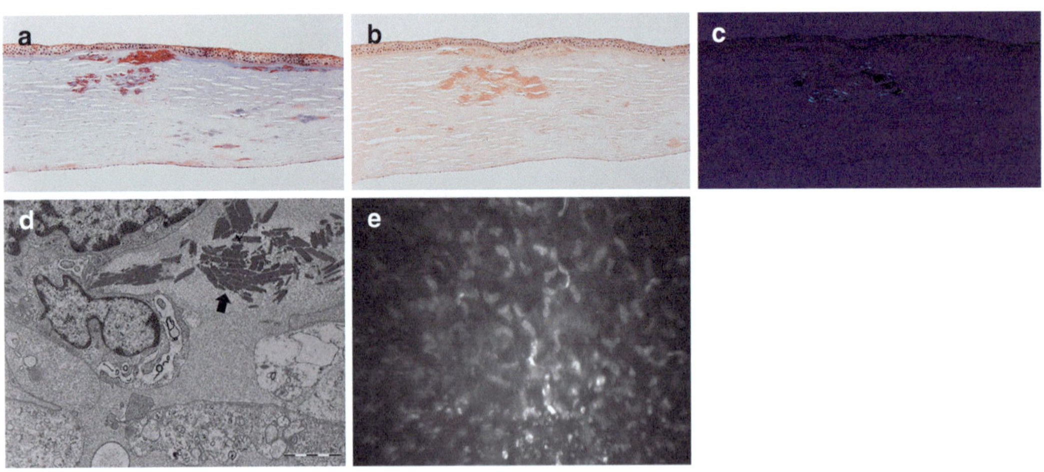

Fig. 7.11 Histopathology of GCD2 cornea. (**a**) Light microscopy—Hyaline deposits in the anterior stroma beneath Bowman's layer are stained red with Masson trichrome stain. (**b, c**) Light microscopy—amyloid deposits in the mid—and deep stroma stain with Congo red (**b**). They also exhibit birefringence and dichroism with cross-polarization (**c**). (**d**) Transmission electron microscopy—keratocytes are enlarged, and intracellular organelles with dilated and degenerative mitochondria and vesicles containing amorphous material are clearly visible. Abnormal deposits are observed (black arrow), and some keratocytes encircle the deposits, extending cellular processes toward the abnormal deposits. (**e**) Confocal microscopy—hyper-reflective irregular deposits are observed in the anterior stroma. Figures (**a–c**) are courtesy of Prof. Hans Grossniklaus, Emory University

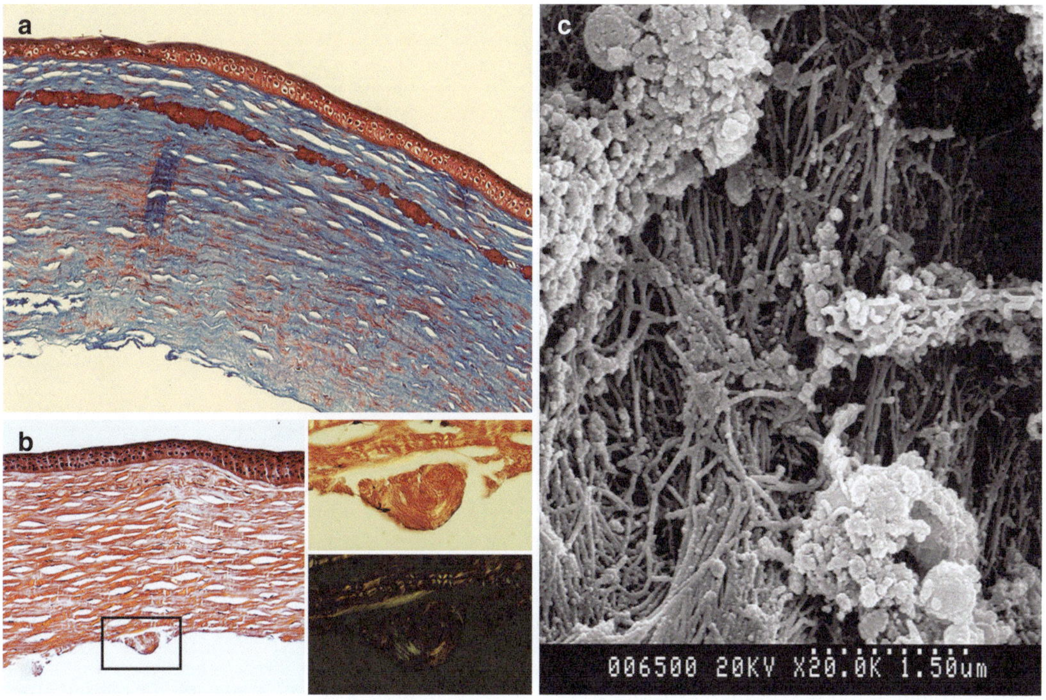

Fig. 7.12 Histopathology of GCD2 cornea exacerbation after LASIK. (**a**) Light microscopy—hyaline deposits mainly located in the LASIK flap interface stain red with Masson trichrome stain. (**b**) Light microscopy—deep stromal lattice-like deposits stain with Congo red and exhibit birefringence, as well as dichroism with cross-polarization (black box). Deposits in the interface did not, however, show birefringence. (**c**) Scanning electron microscopy—small corneal deposits are observed along collagen fibrils, and some clusters of deposits are connected to the collagen fibers

of these deposits is hyaline material (Fig. 7.12) (Roh et al. 2006a; Kim et al. 2008b). Stromal deposits after LASIK stain with antibodies against TGFBIp. Scanning electron microscopy demonstrates many collagen crystal clusters in the flap-bed interface, and abnormal collagen lumps with granular clusters are abundant in the center of the flap (Fig. 7.12) (Roh et al. 2006a). The granular collagen lumps are firmly attached to the cut end of collagen fibers.

Management

Recurrent corneal erosions that may occur in patients with GCD2 should be treated conservatively with artificial tears and contact lenses, as necessary (Han et al. 2013).

Surgical treatment should be considered for cases in which visual acuity decreases and/or contrast sensitivity deteriorates, resulting in loss of functional vision. Various treatment strategies have been used, depending on the depth and extent of corneal lesions, to improve vision by eliminating corneal opacities and delaying further deposition. Phototherapeutic keratectomy (PTK) is the preferred method for removing anterior stromal opacities because it is effective, easily applicable, and repeatable and poses no increased risk of graft failure or rejection (Fig. 7.13).

PTK improves visual acuity and contrast sensitivity, delaying the need for deep anterior lamellar keratoplasty (DALK) or penetrating keratoplasty (PKP). However, recurrence of corneal opacities is often observed after PTK. The reported recurrence rate after PTK varies among published studies from 0% to 100%, depending on the definition of recurrence and the length of the follow-up period (Dinh et al. 1999; Reddy et al. 2013; Chen and Xie 2013; Dogru et al.

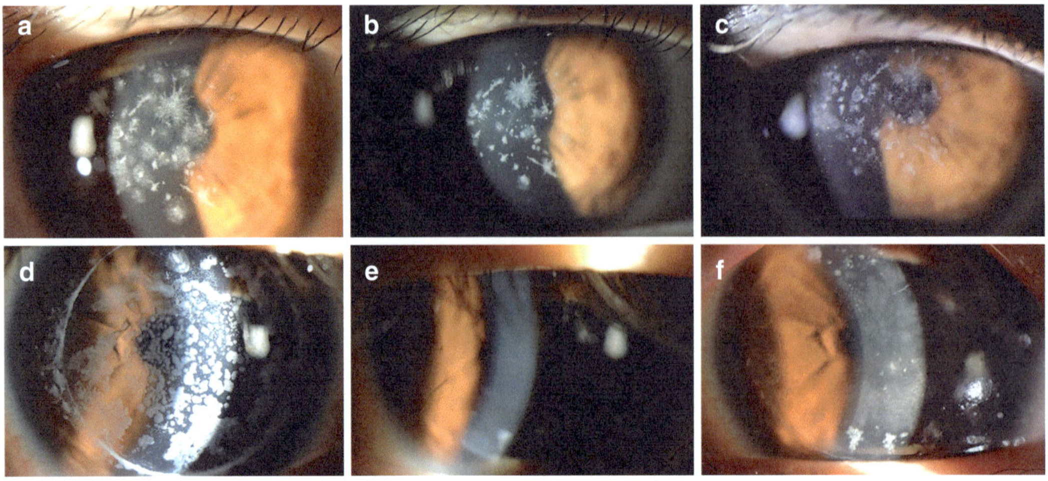

Fig. 7.13 Clinical outcomes of phototherapeutic keratectomy (PTK; **a–c**) and deep anterior lamellar keratoplasty (DALK, **d, e**) in a patient with GCD2. Right eye of a 48-year-old GCD2 heterozygote before PTK (**a**), 3 months (**b**), and 7 years (**c**) after PTK. Recurrence is observed in (**c**). Left eye of a 14-year-old patient with homozygous GCD2 before DALK (**d**), 1 year (**e**), and 4 years (**f**) after DALK. Diffuse superficial haze is observed in the transplanted cornea (**f**)

2001; Orndahl and Fagerholm 1998; Starr et al. 1996; Jun et al. 2018).

Dinh et al. reported that some sign of recurrence was seen in 7 (54%) of 13 eyes with GCD (no subtype mentioned) after PTK, with a mean time to recurrence of 32 months, and significant recurrence (visual loss of 2 lines or more of BCVA, or one or more episodes of recurrent corneal erosion) was noted in 3 (23%) eyes, with a mean time to recurrence of 40 months (Dinh et al. 1999). Dogru et al. reported recurrences between 7 and 9 months in three patients with GCD2 (Dogru et al. 2001). Das et al. found that recurrence of GCD (no subtype mentioned) occurred in 10 (20%) of 50 eyes, at a mean time of 3.4 years (Das et al. 2005). Reddy et al. (2013) reported mild recurrences after PTK in 7/22 (32%) eyes at a mean time of 32 months, and significant recurrences were seen in 4/22 (18%) eyes at a mean time of 47 months after PTK for GCD (no subtype mentioned). Chen and Xie (2013) described significant recurrences an average of 24 months after PTK in granular dystrophy (no subtype mentioned).

In homozygous patients, GCD recurs much more rapidly than it does in heterozygous patients (Moon et al. 2007; Inoue et al. 2002). Inoue et al. reported that the recurrence-free interval was 9.5 months in four eyes of homozygous patients, whereas in seven eyes of heterozygous patients, the recurrence-free interval was 38.4 months (Inoue et al. 2002). Moon et al. reported recurrence of opacities 18 months after first PTK and 3 months after the second or third PTK in a GCD2 homozygote (Moon et al. 2007).

PTK can be repeated in patients who have recurred and can also be applied to patients who have undergone keratoplasty (Maclean et al. 1996). It has been hypothesized that mitomycin C (MMC) might delay the recurrence of GCD2 after PTK (Kim et al. 2006) by preventing the reproduction of keratocytes and, therefore, reducing the production of TGFBIp in the cornea (Kim et al. 2008c). Ha et al., however, studied the effect of MMC during surface ablation with more than 3 years of follow-up and concluded that MMC did not inhibit exacerbation of GCD2 after surface ablation (Ha et al. 2010). The lack of effect of MMC might also be because it also inhibits internalization and degradation of mutant TGFBIp (Choi et al. 2015).

PTK is also effective in patients with an exacerbation of GCD2 after LASIK; however, amputation and removal of the LASIK flap may produce better outcomes than preservation of the flap (Jun et al. 2018). When cataract extraction is indicated in aged eyes of GCD2 heterozygous patients requiring PTK treatment, PTK should be performed first. A formula for calculating the proper intraocular lens implant power after PTK has been suggested by us and others (Jung et al. 2012; Yagi-Yaguchi et al. 2019; Yoneyama et al. 2020).

To treat deep stromal deposits, anterior lamellar keratoplasty (ALKP), DALK, or PKP can be performed. As the corneal endothelium is known not to be affected by *TGFBI*-related corneal dystrophies, ALKP and DALK have become popular surgical treatments. Although there have been few reports of clinical outcomes of keratoplasty in patients with GCD2 compared to other corneal stromal dystrophies, recurrence is often observed even after keratoplasty in patients with GCD2 (Fig. 7.13). Ellies et al. described two eyes with GCD2 that underwent PKP and did not show significant recurrence during the follow-up period (Ellies et al. 2002). Ünal et al. reported that the recurrence rate after DALK was 22% (two eyes) in nine eyes with GCD (no subtype mentioned) after a mean time of 33 months (Unal et al. 2013). Park et al. found that 1 (25%) of four eyes with GCD2 showed recurrence with small opacities in the peripheral stroma 13 months after DALK, and other eyes showed no recurrence during 14 to 21 months after DALK (Park et al. 2007). Salouti et al. reported that mild recurrence was noted in 5 (71%) eyes with a mean time of 15.6 months, and clinically significant recurrence occurred in 3 (43%) eyes with mean time of 34 months after DALK using the Melles DALK technique for GCD (no subtype mentioned) (Salouti et al. 2009).

Since many previous studies suggest that corneal epithelial cells are the origin of deposits in *TGFBI*-related corneal dystrophies, limbal stem cell transplantation has been tried as a treatment (Sundmacher et al. 1999; Dunaief et al. 2001; Spelsberg et al. 2004; Lang et al. 2015). Although several reports have suggested that this surgical procedure may be an effective for treatment for corneal stromal dystrophies, it is not widely used because rejection of transplanted limbal stem cells remains problematic.

Kamoi et al. (2005) demonstrated that corneal electrolysis could remove deposits from two homozygotic eyes with recurrent GCD2 cornea after keratoplasty. In these eyes, fine subepithelial opacities began to appear 3 years after the treatment. The long-term clinical outcome of this method remains unknown.

It has been proposed that treatment of corneal dystrophies be based on the depth of opacities (Seitz and Lisch 2011). Initially, conservative therapy such as topical medications and contact lenses can be applied. If these treatments are ineffective, PTK, which can be performed multiple times, might postpone the need for corneal transplantation. Eventually PTK might no longer be applicable because of corneal thinning, scar formation, or an undesirable corneal curvature. Fourier-domain optical coherence tomography (FD-OCT) can be used as an aid to slit lamp examination to determine the depth and extent of lesions to determine the depth of PTK required for removal of opacities (Kim et al. 2010). Jung et al. (2013) described the use of preoperative FD-OCT and intraoperative slit lamp examination to minimize the ablation depth of PTK required for adequate removal of opacities (Fig. 7.14). They removed diffuse haze from the superficial stroma, which is the main cause of the visual impairment in advanced stages of disease. The mean depth of PTK ablation was 43.7 μm (range, 31–59 μm) in 43 eyes of GCD2 patients, and no recurrence was observed during a mean follow-up period of 21 months.

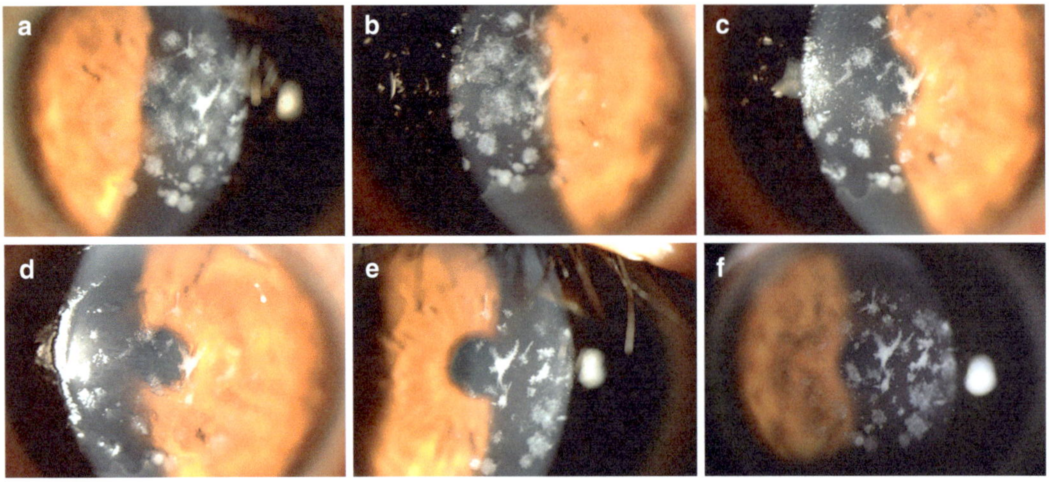

Fig. 7.14 Serial photographs taken before and during PTK for diffuse stromal haze in a patient with GCD2. The preoperative slit-lamp photograph of the left eye of a 60-year-old heterozygous GCD2 patient shows a few discrete large and small anterior stromal opacities with diffuse stromal haze between opacities (**a**). Intraoperative slit-lamp photographs taken after PTK at ablation depths of 22 (**b**), 34 (**c**), and 41 μm (**d**), respectively. Slit-lamp photographs taken 2 months (**e**) and 10 years (**f**) after PTK

References

Abazi Z, et al. Individual phenotypic variances in a family with Avellino corneal dystrophy. BMC Ophthalmol. 2013;13:30.

Afshari NA, et al. Survey of patients with granular, lattice, avellino, and Reis-Bucklers corneal dystrophies for mutations in the BIGH3 and gelsolin genes. Arch Ophthalmol. 2001;119(1):16–22.

Akiya S, et al. Granular-lattice (Avellino) corneal dystrophy. Ophthalmologica. 1999;213(1):58–62.

Aldave AJ, et al. Lattice corneal dystrophy associated with the Ala546Asp and Pro551Gln missense changes in the TGFBI gene. Am J Ophthalmol. 2004;138(5):772–81.

Aldave AJ, et al. A clinical and histopathologic examination of accelerated TGFBIp deposition after LASIK in combined granular-lattice corneal dystrophy. Am J Ophthalmol. 2007;143(3):416–9.

Aldave AJ, et al. A novel variant of combined granular-lattice corneal dystrophy associated with the Met619Lys mutation in the TGFBI gene. Arch Ophthalmol. 2008;126(3):371–7.

Ann LB, et al. Variant lattice corneal dystrophy associated with compound heterozygous mutations in the TGFBI gene. Br J Ophthalmol. 2017;101(4):509–13.

Banning CS, et al. Exacerbation of Avellino corneal dystrophy after LASIK in North America. Cornea. 2006;25(4):482–4.

Bouyacoub Y, et al. Granular type I corneal dystrophy in a large consanguineous Tunisian family with homozygous p.R124S mutation in the TGFBI gene. Ophthalmic Genet. 2019;40(4):329–37.

Chae H, et al. Mutational spectrum of Korean patients with corneal dystrophy. Clin Genet. 2016;89(6):678–89.

Chen M, Xie L. Features of recurrence after excimer laser phototherapeutic keratectomy for anterior corneal pathologies in North China. Ophthalmology. 2013;120(6):1179–85.

Cho KJ, et al. TGFBI gene mutations in a Korean population with corneal dystrophy. Mol Vis. 2012;18:2012–21.

Choi SI, et al. Impaired autophagy and delayed autophagic clearance of transforming growth factor β-induced protein (TGFBI) in granular corneal dystrophy type 2. Autophagy. 2012;8(12):1782–97.

Choi SI, et al. Lysosomal trafficking of TGFBIp via caveolae-mediated endocytosis. PLoS One. 2015;10(4):e0119561.

Das S, Langenbucher A, Seitz B. Excimer laser phototherapeutic keratectomy for granular and lattice corneal dystrophy: a comparative study. J Refract Surg. 2005;21(6):727–31.

Dighiero P, et al. A novel variant of granular corneal dystrophy caused by association of 2 mutations in the TGFBI gene-R124L and DeltaT125-DeltaE126. Arch Ophthalmol. 2000;118(6):814–8.

Dinh R, et al. Recurrence of corneal dystrophy after excimer laser phototherapeutic keratectomy. Ophthalmology. 1999;106(8):1490–7.

Dogru M, et al. Alteration of the ocular surface with recurrence of granular/avellino corneal dystrophy after phototherapeutic keratectomy: report of five cases and literature review. Ophthalmology. 2001;108(4):810–7.

Dunaief JL, Ng EW, Goldberg MF. Corneal dystrophies of epithelial genesis: the possible therapeutic use of limbal stem cell transplantation. Arch Ophthalmol. 2001;119(1):120–2.

Ellies P, et al. Clinical outcome of eight BIGH3-linked corneal dystrophies. Ophthalmology. 2002;109(4):793–7.

Folberg R, et al. Clinically atypical granular corneal dystrophy with pathologic features of lattice-like amyloid deposits. A study of these families. Ophthalmology. 1988;95(1):46–51.

Frising M, et al. Recurrent granular dystrophy of the cornea: an unusual case. Cornea. 2006;25(5):614–7.

Gruenauer-Kloevekorn C, et al. TGFBI (BIGH3) gene mutations in German families: two novel mutations associated with unique clinical and histopathological findings. Br J Ophthalmol. 2009;93(7):932–7.

Ha BJ, et al. Mitomycin C does not inhibit exacerbation of granular corneal dystrophy type II induced by refractive surface ablation. Cornea. 2010;29(5):490–6.

Han KE, et al. Clinical findings and treatments of granular corneal dystrophy type 2 (avellino corneal dystrophy): a review of the literature. Eye Contact Lens. 2010;36(5):296–9.

Han KE, et al. Extremely varied phenotypes in granular corneal dystrophy type 2 heterozygotes. Mol Vis. 2012;18:1755–62.

Han KE, et al. Changes of clinical manifestation of granular corneal deposits because of recurrent corneal erosion in granular corneal dystrophy types 1 and 2. Cornea. 2013;32(5):e113–20.

Han KE, et al. Pathogenesis and treatments of TGFBI corneal dystrophies. Prog Retin Eye Res. 2016;50:67–88.

Holland EJ, et al. Avellino corneal dystrophy. Clinical manifestations and natural history. Ophthalmology. 1992;99(10):1564–8.

Inoue T, et al. Recurrence of corneal dystrophy resulting from an R124H Big-h3 mutation after phototherapeutic keratectomy. Cornea. 2002;21(6):570–3.

Jun RM, et al. Avellino corneal dystrophy after LASIK. Ophthalmology. 2004;111(3):463–8.

Jun I, et al. Long-term clinical outcomes of phototherapeutic keratectomy in corneas with granular corneal dystrophy type 2 exacerbated after LASIK. J Refract Surg. 2018;34(2):132–9.

Jung SH, et al. Intraocular lens power calculations for cataract surgery after phototherapeutic keratectomy in granular corneal dystrophy type 2. J Refract Surg. 2012;28(10):714–24.

Jung SH, et al. Phototherapeutic keratectomy in diffuse stromal haze in granular corneal dystrophy type 2. Cornea. 2013;32(3):296–300.

Kamoi M, et al. Electrolysis for corneal opacities in a young patient with superficial variant of granular corneal dystrophy (Reis-Bücklers corneal dystrophy). Am J Ophthalmol. 2005;139(6):1139–40.

Kim HS, et al. BIGH3 gene mutations and rapid detection in Korean patients with corneal dystrophy. Cornea. 2001;20(8):844–9.

Kim TI, et al. Mitomycin C inhibits recurrent Avellino dystrophy after phototherapeutic keratectomy. Cornea. 2006;25(2):220–3.

Kim TI, et al. Comparison of corneal deposits after LASIK and PRK in eyes with granular corneal dystrophy type II. J Refract Surg. 2008a;24(4):392–5.

Kim TI, et al. Deposits of transforming growth factor-beta-induced protein in granular corneal dystrophy type II after LASIK. Cornea. 2008b;27(1):28–32.

Kim TI, et al. Mitomycin C induces apoptosis in cultured corneal fibroblasts derived from type II granular corneal dystrophy corneas. Mol Vis. 2008c;14:1222–8.

Kim TI, et al. Determination of treatment strategies for granular corneal dystrophy type 2 using Fourier-domain optical coherence tomography. Br J Ophthalmol. 2010;94(3):341–5.

Klintworth GK, Bao W, Afshari NA. Two mutations in the TGFBI (BIGH3) gene associated with lattice corneal dystrophy in an extensively studied family. Invest Ophthalmol Vis Sci. 2004;45(5):1382–8.

Korvatska E, et al. Mutation hot spots in 5q31-linked corneal dystrophies. Am J Hum Genet. 1998;62(2):320–4.

Kwak JJ, et al. Exacerbation of granular corneal dystrophy type 2 after small incision lenticule extraction. Cornea. 2021;40(4):519–24.

Lakshminarayanan R, et al. A novel mutation in transforming growth factor-beta induced protein (TGFβIp) reveals secondary structure perturbation in lattice corneal dystrophy. Br J Ophthalmol. 2011;95(10):1457–62.

Lang SJ, et al. Simultaneous transplantation of limbal stem cells may reduce recurrences of granular dystrophy after corneal transplantation: 2 long-term case reports. Medicine (Baltimore). 2015;94(20):e789.

Lee JH, et al. Effects of corneal neovascularization on the manifestations of Avellino corneal dystrophy (granular corneal dystrophy type II). Cornea. 2006;25(8):914–8.

Lee JH, et al. Exacerbation of granular corneal dystrophy type II (Avellino corneal dystrophy) after LASEK. J Refract Surg. 2008;24(1):39–45.

Lee JH, et al. Prevalence of granular corneal dystrophy type 2 (Avellino corneal dystrophy) in the Korean population. Ophthalmic Epidemiol. 2010;17(3):160–5.

Lisch W, Weiss JS. Clinical and genetic update of corneal dystrophies. Exp Eye Res. 2019;186:107715.

Maclean H, et al. Excimer phototherapeutic keratectomy for recurrent granular dystrophy. Aust N Z J Ophthalmol. 1996;24(2):127–30.

Mashima Y, et al. Severe form of juvenile corneal stromal dystrophy with homozygous R124H mutation in the keratoepithelin gene in five Japanese patients. Br J Ophthalmol. 1998;82(11):1280–4.

Mashima Y, et al. Association of autosomal dominantly inherited corneal dystrophies with BIGH3 gene mutations in Japan. Am J Ophthalmol. 2000;130(4):516–7.

Moon JW, et al. Homozygous granular corneal dystrophy type II (Avellino corneal dystrophy): natural history and progression after treatment. Cornea. 2007;26(9):1095–100.

Munier FL, et al. Kerato-epithelin mutations in four 5q31-linked corneal dystrophies. Nat Genet. 1997;15(3):247–51.

Musch DC, et al. Prevalence of corneal dystrophies in the United States: estimates from claims data. Invest Ophthalmol Vis Sci. 2011;52(9):6959–63.

Niel-Butschi F, et al. Genotype-phenotype correlations of TGFBI p.Leu509Pro, p.Leu509Arg, p.Val613Gly, and the allelic association of p.Met502Val-p.Arg555Gln mutations. Mol Vis. 2011;17:1192–202.

Nowinska AK, et al. Genotype-phenotype correlation of TGFBI corneal dystrophies in Polish patients. Mol Vis. 2011;17:2333–42.

Okada M, et al. Severe corneal dystrophy phenotype caused by homozygous R124H keratoepithelin mutations. Invest Ophthalmol Vis Sci. 1998;39(10):1947–53.

Orndahl MJ, Fagerholm PP. Treatment of corneal dystrophies with phototherapeutic keratectomy. J Refract Surg. 1998;14(2):129–35.

Park KA, et al. Deep anterior lamellar keratoplasty in Korean patients with Avellino dystrophy. Cornea. 2007;26(9):1132–5.

Reddy JC, et al. Excimer laser phototherapeutic keratectomy in eyes with corneal stromal dystrophies with and without a corneal graft. Am J Ophthalmol. 2013;155(6):1111–1118 e2.

Roh MI, et al. Avellino corneal dystrophy exacerbated after LASIK: scanning electron microscopic findings. Cornea. 2006a;25(3):306–11.

Roh MI, et al. Preserved peripheral corneal clarity after surgical trauma in patients with Avellino corneal dystrophy. Cornea. 2006b;25(4):497–8.

Rosenwasser GO, et al. Phenotypic variation in combined granular-lattice (Avellino) corneal dystrophy. Arch Ophthalmol. 1993;111(11):1546–52.

Sakimoto T, et al. A novel nonsense mutation with a compound heterozygous mutation in TGFBI gene in lattice corneal dystrophy type I. Jpn J Ophthalmol. 2003;47(1):13–7.

Salouti R, et al. Deep anterior lamellar keratoplasty with melles technique for granular corneal dystrophy. Cornea. 2009;28(2):140–3.

Seitz B, Lisch W. Stage-related therapy of corneal dystrophies. Dev Ophthalmol. 2011;48:116–53.

Song JS, et al. Mutation analysis of the TGFBI gene in consecutive korean patients with corneal dystrophies. Ann Lab Med. 2015;35(3):336–40.

Spelsberg H, et al. Penetrating limbo-keratoplasty for granular and lattice corneal dystrophy: survival of donor limbal stem cells and intermediate-term clinical results. Ophthalmology. 2004;111(8):1528–33.

Starr M, et al. Excimer laser phototherapeutic keratectomy. Cornea. 1996;15(6):557–65.

Stewart HS, et al. Heterogeneity in granular corneal dystrophy: identification of three causative mutations in the TGFBI (BIGH3) gene-lessons for corneal amyloidogenesis. Hum Mutat. 1999;14(2):126–32.

Streeten BW, et al. Immunolocalization of beta ig-h3 protein in 5q31-linked corneal dystrophies and normal corneas. Arch Ophthalmol. 1999;117(1):67–75.

Sundmacher R, Spelsberg H, Reinhard T. Homologous penetrating central limbokeratoplasty in granular and lattice corneal dystrophy. Cornea. 1999;18(6):664–70.

Tsujikawa K, et al. Allelic homogeneity in Avellino corneal dystrophy due to a founder effect. J Hum Genet. 2007;52(1):92–7.

Unal M, et al. Deep anterior lamellar keratoplasty for the treatment of stromal corneal dystrophies. Cornea. 2013;32(3):301–5.

Wan XH, et al. Exacerbation of Avellino corneal dystrophy after laser in situ keratomileusis. Cornea. 2002;21(2):223–6.

Watanabe H, et al. Two patterns of opacity in corneal dystrophy caused by the homozygous BIG-H3 R124H mutation. Am J Ophthalmol. 2001;132(2):211–6.

Weiss JS, et al. The IC3D classification of the corneal dystrophies. Cornea. 2008;27 Suppl 2(Suppl 2):S1–83.

Weiss JS, et al. IC3D classification of corneal dystrophies—edition 2. Cornea. 2015;34(2):117–59.

Yagi-Yaguchi Y, et al. Comparison of the accuracy of newer intraocular lens power calculation methods in eyes that underwent previous phototherapeutic keratectomy. J Refract Surg. 2019;35(5):310–6.

Yam GH, et al. In vitro amyloid aggregate forming ability of TGFBI mutants that cause corneal dystrophies. Invest Ophthalmol Vis Sci. 2012;53(9):5890–8.

Yamada N, et al. Double mutation (R124H, N544S) of TGFBI in two sisters with combined expression of Avellino and lattice corneal dystrophies. Mol Vis. 2009;15:974–9.

Yamamoto S, et al. The spectrum of beta ig-h3 gene mutations in Japanese patients with corneal dystrophy. Cornea. 2000;19(3 Suppl):S21–3.

Yoneyama R, et al. Predictability of intraocular lens power calculation in eyes after phototherapeutic keratectomy. Jpn J Ophthalmol. 2020;64(1):62–7.

Yu J, et al. Analysis of mutation of BIGH3 gene in Chinese patients with corneal dystrophies. Zhonghua Yan Ke Za Zhi. 2003;39(10):582–6.

Zenteno JC, et al. Clinical and genetic features of TGFBI-linked corneal dystrophies in Mexican population: description of novel mutations and novel genotype-phenotype correlations. Exp Eye Res. 2009;89(2):172–7.

Zhong X, et al. Novel and known mutations of TGFBI, their genotype-phenotype correlation and structural modeling in 3 Chinese families with lattice corneal dystrophy. Mol Vis. 2010;16:224–30.

Molecular Pathogenesis of Granular Corneal Dystrophy Type 2 (GCD2) and *TGFBI*-Related Corneal Dystrophies

8

Ikhyun Jun, Seung-il Choi, Tae-im Kim, Hyung Keun Lee, and Eung Kweon Kim

Overview of *TGFBI* Gene

The transforming growth factor beta-induced gene (*TGFBI*; *BIGH3*; *βigh3*) was first discovered in a human lung adenocarcinoma cell line, where it was upregulated after treatment with the pleiotropic cytokine TGF-β1 (Skonier et al. 1992). *TGFBI* mRNA encodes a 68-kDa extracellular matrix (ECM) protein, transforming growth factor beta-induced protein (TGFBIp), also described as MP78/70 (Gibson et al. 1997), collagen fiber-associated protein (RGD-CAP) (Hashimoto et al. 1997), BIGH3, βigh3, or kerato-epithelin (Munier et al. 1997; Han et al. 2016). TGFBIp consists of an N-terminal 23 residues of secretory signal peptide sequence, a cysteine-rich EMI domain, four homologous fasciclin 1 (FAS1) domains which each contain 140 amino acid residues at the N-terminus, and an arginine-glycine-aspartate (RGD) motif which binds to integrin at the C-terminus (Han et al. 2016). The cysteine-rich EMI domain contains 6 of the 11 cysteine residues and has been revealed as an EMI domain because of sequence homology (Doliana et al. 2000).

The various functional roles of TGFBIp have been identified and are still under investigation. TGFBIp plays an important role in physiologic and pathologic reactions by regulating the proliferation, adhesion, migration, and differentiation (Skonier et al. 1992; Nam et al. 2005; Park et al. 2004, 2009). TGFBIp has been reported to control cellular adhesion and migration through the interaction with various integrins (Kim and Kim 2008; Kim et al. 2002, 2003; Nam et al.

I. Jun · T.-i. Kim
Department of Ophthalmology, Yonsei University College of Medicine, Seoul, Korea (Republic of)

The Corneal Dystrophy Research Institute, Yonsei University College of Medicine, Seoul, Korea (Republic of)

Instutute of Vision Research, Yonsei University College of Medicine, Seoul, Korea (Republic of)
e-mail: HADESDUAL@yuhs.ac; TIKIM@yuhs.ac

S.-i. Choi
The Corneal Dystrophy Research Institute, Yonsei University College of Medicine, Seoul, Korea (Republic of)

Department of Food Marketing and Safety, Konkuk University, Seoul, Korea (Republic of)
e-mail: seung1choi@konkuk.ac.kr

H. K. Lee
Department of Ophthalmology, Yonsei University College of Medicine, Seoul, Korea (Republic of)

Instutute of Vision Research, Yonsei University College of Medicine, Seoul, Korea (Republic of)
e-mail: shadik@yuhs.ac

E. K. Kim (✉)
Department of Ophthalmology, Yonsei University College of Medicine, Seoul, Korea (Republic of)

The Corneal Dystrophy Research Institute, Yonsei University College of Medicine, Seoul, Korea (Republic of)

Saevit Eye Hospital, Goyang, Korea (Republic of)
e-mail: eungkkim@yuhs.ac

E. K. Kim, R. D. Stulting (eds.), *TGFBI-related Corneal Dystrophies*,
https://doi.org/10.1007/978-981-96-0131-8_8

2006; Ohno et al. 1999). TGFBIp also mediates progression, suppression, dissemination, and metastasis of malignant tumors (Nam et al. 2005; Fan et al. 2014; Massague 2008). A recent study has revealed that TGFBIp increases adhesion, migration, and differentiation of human lymphatic endothelial cells; hence suppression of TGFBIp reduced lymphangiogenesis of tumor and metastasis of TGFBIp-producing tumors (Maeng et al. 2016). The localized expression of TGFBIp at wounds and tumors augments the recruitment of endothelial progenitor cells and regulates the expression of endothelial genes through integrins α4 and α5 (Maeng et al. 2015a).

TGFBIp is known to be expressed ubiquitously in various organs and tissues including the heart, breast, intestines, liver, pancreas, lung, kidney, skin, testes, prostate, ovary, endometrium, bone, tendon, and blood plasma (Skonier et al. 1992; Klintworth et al. 1998; Kitahama et al. 2000; Ferguson et al. 2003a, b; Carson et al. 2002; Lee et al. 2003). Also, TGFBIp is one of the major proteins in the human cornea (Klintworth et al. 1994). TGFBIp exists in all layers of the human cornea and is the second most abundant in the corneal stroma (Dyrlund et al. 2012). It is known that the majority of corneal TGFBIp is expressed in the epithelium (Escribano et al. 1994). A recent study found that *TGFBI* mRNA expression was limited to the corneal epithelium by in situ hybridization experiment in mice (Poulsen et al. 2018). In the human, however, TGFBIp is also expressed in keratocytes, especially in the peripheral cornea (Liu et al. 2017). TGFBIp was detected in the subepithelial matrix, posterior collagenous layer, and abnormal Descemet's membrane co-localized with type VI collagen in Fuchs' dystrophy (Hirano et al. 1996). Approximately, 60% of corneal TGFBIp is covalently linked to type XII collagen (Runager et al. 2013), and the expression of TGFBIp is greatly increased during wound healing in the rabbit cornea (Rawe et al. 1997). Recently, TGFBIp has been shown to increase migration, proliferation, adhesion, and wound healing in immortalized human corneal epithelial cells (Maeng et al. 2017). In addition, TGFBIp augments the expression of mucins including MUC-1, MUC-4, MUC-5 AC, and MUC-16 in human corneal and conjunctival epithelial cells via the ERK and AKT signaling pathways (Maeng et al. 2017). TGFBIp is thought to play pivotal roles in wound healing and maintenance of the ECM in the cornea.

Morphological Characteristics of Corneal Fibroblasts in GCD2 and *TGFBI*-Related Corneal Dystrophies

Keratocytes, called corneal fibroblasts when cultured, are the major cellular component of the corneal stroma and occupy 9–17% of stroma volume (Hahnel et al. 2000). The human cornea was found to have a keratocyte density of about 23,000 cells/mm^3 at mid-stroma (Hahnel et al. 2000; McLaren et al. 2005). Keratocytes serve to maintain corneal transparency and its ECM (Jester et al. 1999). The morphological changes in corneal fibroblasts from patients with *TGFBI*-related corneal dystrophies have been studied primarily in patients with GCD2. As a result, this chapter will focus on the changes in granular corneal dystrophy type 2 (GCD2) with p.(Arg124His) mutation.

Compared to wild-type corneal fibroblasts, corneal fibroblasts in GCD2 are large in size; have numerous, variably sized vesicles that contain an amorphous substance; and show a senescent morphology (Fig. 8.1). TGFBIp co-localizes with the lysosomal enzyme, cathepsin D. Homozygous GCD2 corneal fibroblasts present a more extensive co-localization pattern between TGFBIp and cathepsin D than that shown in wild-type corneal fibroblasts (Choi et al. 2012). In addition, mitochondria in GCD2 showed fragmented and swollen features compared with the small, round mitochondria seen in normal corneal fibroblasts by transmission electron microscopy (Kim et al. 2011). The proliferation rates of corneal fibroblasts were significantly lower in GCD2 homozygotes than in normal (Choi et al. 2015a). The reduced number of elongated and disorganized mitochondria was notably observed in late passages of homozygous GCD2 corneal fibroblasts. The corneal fibroblasts from homozygous GCD2 patients do not survive more than eight passages,

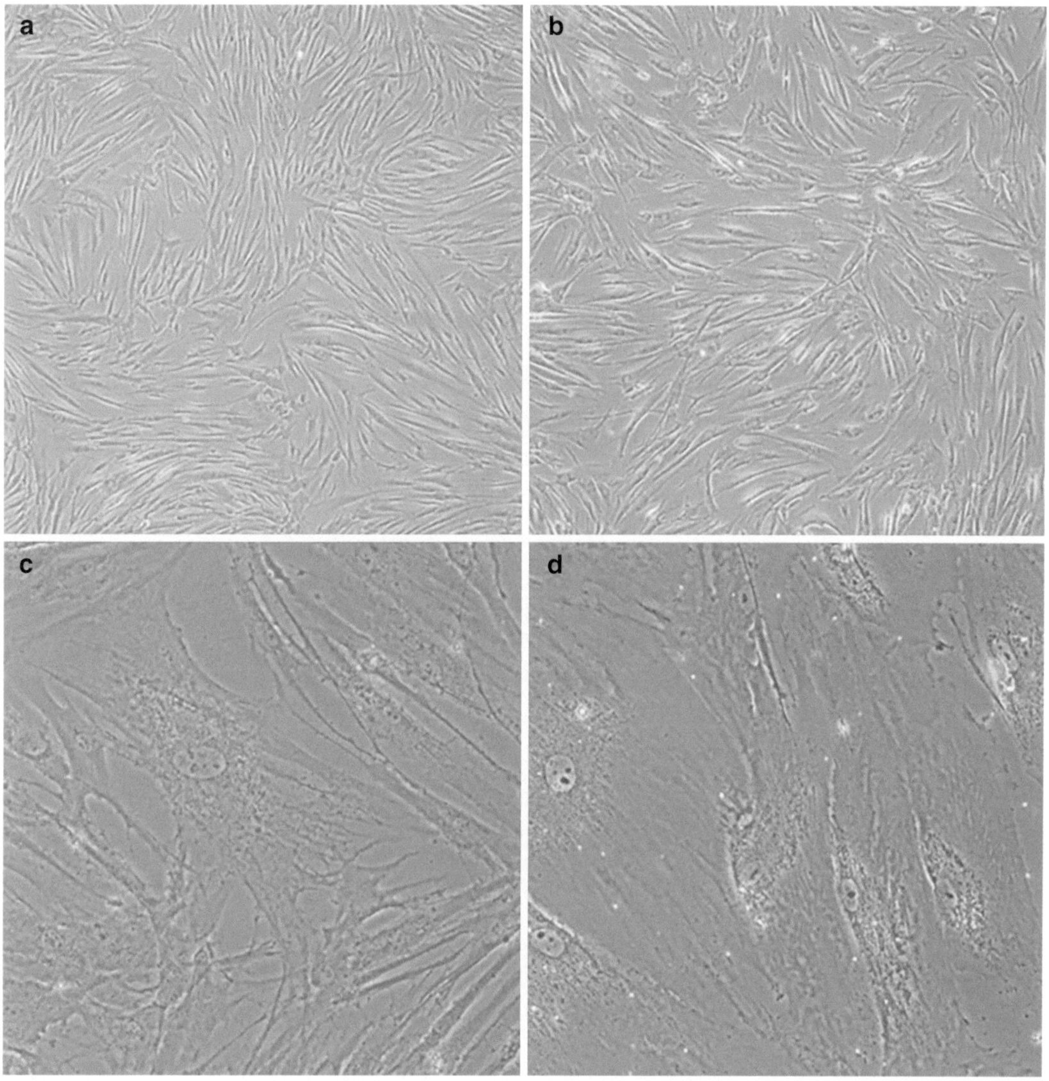

Fig. 8.1 Morphology of primary cultured corneal fibroblasts of wild-type (**a, c**) and GCD2 homozygotes (**b, d**) observed by phase-contrast microscopy. The homozygous GCD2 corneal fibroblasts are increased in size and have intracellular vesicles with various sizes. Reprinted from (Han et al. 2016) with permission from Elsevier

while normal cells can survive more than 24 culture passages in vitro.

At least a twofold difference in expression of 555 genes were detected between wild type and homozygous GCD2 primary cultured corneal fibroblasts by microarray assay (Choi et al. 2010). Among these, 319 genes showed an increased expression, while 236 genes showed a decreased expression in the GCD2 corneal fibroblasts compared to wild-type corneal fibroblasts. These genes are known to be associated with various intracellular signal transduction and physiological functions, involving the TGF-β receptor- and integrin-mediated signaling, cell differentiation and proliferation, immune responses, cell adhesion, ECM proteolytic enzymes, cell cycle, cytoskeletal organization, mitochondrial energy metabolism, collagen catabolism, response to wounding, response to oxidative stress, and the ubiquitin-mediated proteasomal degradation pathway (Choi et al. 2010). The complexity of these gene functions can lead to cellular changes in corneal fibroblasts of *TGFBI*-related dystrophies, and further research is needed to determine whether new factors are involved in pathogenesis of *TGFBI*-related dystrophies.

Oxidative Stress in *TGFBI*-Related Corneal Dystrophies

Reactive oxygen species (ROS), which can cause oxidative damage to DNA, proteins, and lipids, are generated in mitochondria and peroxisomes during oxygen-based energy metabolism in mammalian cells (Kim et al. 2015). The cornea absorbs a large amount of ultraviolet (UV) light that would otherwise enter the eye. It is well known that UV exposure generates ROS and causes tissue damage (Wenk et al. 2001). During the oxidative phosphorylation reaction, which generates ATP, toxic oxygen-derived free radicals, such as superoxide (O_2^-), hydroxyl radicals (•OH), and hydrogen peroxide (H_2O_2), are produced in mitochondria (Mates et al. 1999). ROS can react with cellular proteins, nucleic acids, and lipids and release by-products, including protein carbonyl groups, 8-hydroxy-2′-deoxyguanosin (8-OHdG), malondialdehyde (MDA), and 4-hydroxy-2-nonenal (HNE), which are the signs of oxidative damage (Han et al. 2016). To minimize oxidative damage, mammalian cells have developed several ROS-scavenging mechanisms, including superoxide dismutase (SOD), glutathione peroxidase, glutathione reductase, and non-enzymatic ROS scavenging systems, to reduce the concentration of ROS (Rao et al. 1987).

The expression levels of the MDA, 4-HNE, and protein carbonyl groups are significantly increased in GCD2 corneal fibroblasts compared to normal cells (Choi et al. 2009). Immunoreactivity of MDA is stronger in the corneal epithelium and stroma of GCD2 patients' corneas than that of age-matched wild-type corneas (Fig. 8.2). Intracellularly generated ROS and H_2O_2 are increased significantly in GCD2, and expression of antioxidant enzymes, such as Cu/Zn-SOD, Mn-SOD, glutathione peroxidase, and glutathione reductase, is significantly elevated in GCD2 corneal fibroblasts, compared to those of wild-type cells. Interestingly, the protein level of catalase is significantly lower in GCD2 corneal fibroblasts, although mRNA expression level of catalase is significantly higher in GCD2 than in wild-type corneal fibroblasts (Choi et al. 2009). Immunoreactivity of catalase is lower, and activity of catalase is also significantly weaker in corneal fibroblasts from GCD2 (Fig. 8.3). Corneal fibroblasts from GCD2 patients are more susceptible to H_2O_2-induced cell death, and the expression of anti-apoptotic proteins, Bcl-2, and Bcl-xL is significantly decreased; whereas, Bax, and Bok, pro-apoptotic proteins are significantly induced in GCD2, compared to those in wild-type corneal fibroblasts (Choi et al. 2009). A significant upregulation of the endoplasmic reticulum (ER) stress markers including binding immunoglobulin protein (BiP), protein disulfide isomerase (PDI), double-stranded RNA-activated protein kinase (PKR)-like ER kinase (PERK), and inositol requiring enzyme 1α (IRE1α) were noted in GCD2 corneal fibroblasts (Choi et al. 2016a). Since XBP1 mRNA splicing is mediated by IRE1α, XBP1s protein level is significantly augmented in GCD2 corneal fibroblasts. In summary, the corneal fibroblasts of patients with *TGFBI*-related corneal dystrophies have more oxidative stress and are more vulnerable to oxidative injury.

Recently, 4-phenylbutyric acid (4-PBA) has been reported to have chaperone activity and reduce ER stress and cell death by reversing the misfolding of many disease-related proteins (Choi et al. 2016a). Treatment of 4-PBA significantly attenuated upregulation of BiP, IRE1α, and XBP1s by tunicamycin and decreased TGFBIp expression in corneal fibroblasts (Choi et al. 2016a). However, 4-PBA did not have an influence on *TGFBI* mRNA levels and LC3 II levels. Induction of TGFBIp level by co-treatment with tunicamycin and MG132, or bafilomycin A1, was suppressed by 4-PBA. Taken together, these data suggest that 4-PBA might be a potential therapy for *TGFBI*-related corneal dystrophy.

Melatonin reduced ER stress-induced cell death by tunicamycin in both wild-type and GCD2 corneal fibroblasts (Choi et al. 2017). Tunicamycin-induced cellular shrinkage,

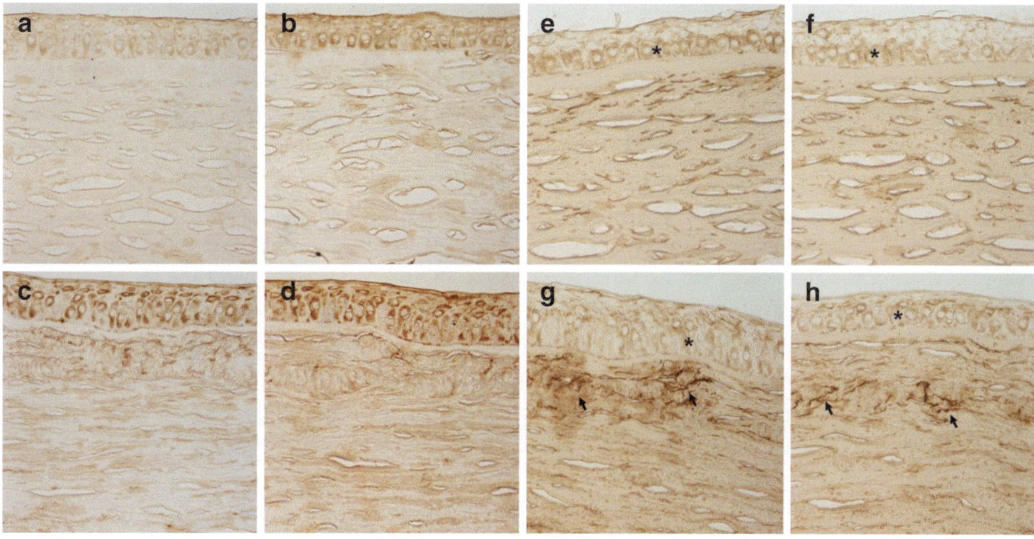

Fig. 8.2 In a normal cornea (**a, b**), MDA immunoreactivity was present in the cytoplasmic compartments of epithelial cells. The corneal stroma also displayed diffused MDA immunoreactivity. In a GCD2 cornea (**c, d**), strong MDA immunoreactivity was present in the cytoplasmic compartments of most epithelial cells as well as the stromal matrix. The immunohistochemical staining for TGFBIp is most prominent in epithelial cells (asterisk). TGFBIp deposits (arrows) are shown on the stroma of corneal tissue (**g, h**): normal (**e, f**) and homozygote cornea (**g, h**). Reprinted from (Choi et al. 2009) with permission from Elsevier

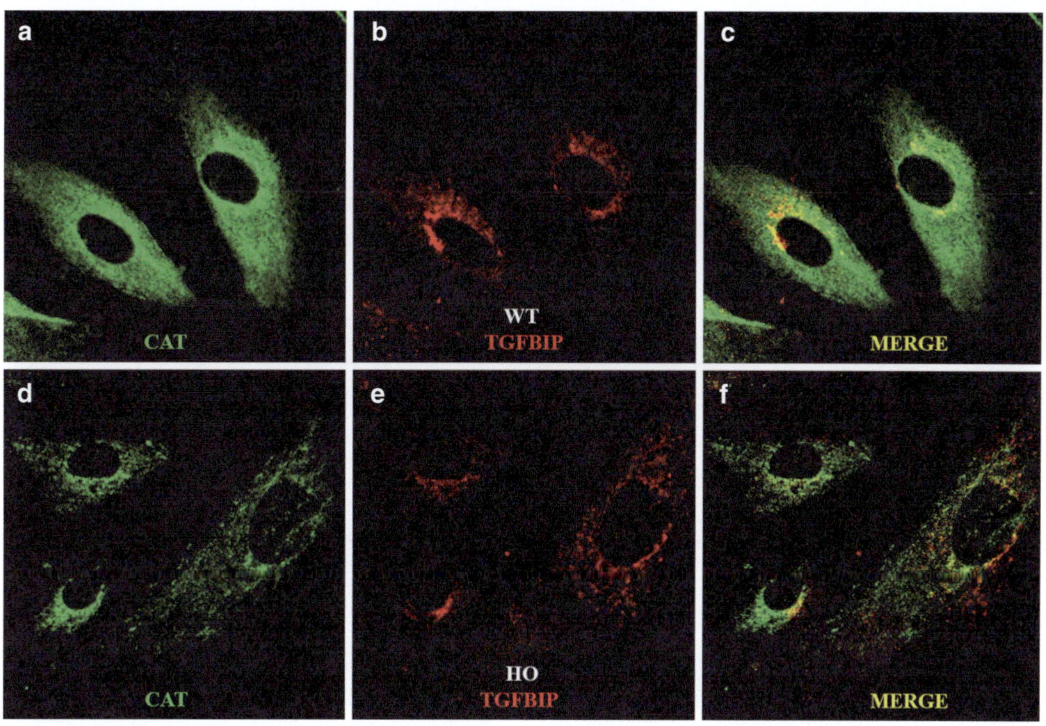

Fig. 8.3 (**a–f**) The subcellular localization of catalase and TGFBIp in homozygous and wild-type primary cultured corneal fibroblasts (PCFs). Immunostaining for catalase was less intense in homozygous (**d**) than wild-type (**a**) PCFs. Magnification = X60. *CAT* catalase. Reprinted from (Choi et al. 2009) with permission from Elsevier

reduction of attachment rate, and apoptotic death were rescued by melatonin treatment of GCD2 corneal fibroblasts. Upregulation of BiP, IRE1α, PERK, and ATF6 by ER stress was suppressed by melatonin in wild-type and GCD2 corneal fibroblasts. Also, XBP1 splicing and TGFBIp expression were inhibited by melatonin. These effects of melatonin were reduced by MG132, a ubiquitin-proteasome inhibitor, not by bafilomycin A1, an autophagy inhibitor, suggesting that ER-associated protein degradation (ERAD) is involved. Melatonin promoted SEL1L, ERAD system component, which is reduced in GCD2 cells (Choi et al. 2017). Melatonin may also have potential as a treatment option for *TGFBI*-related corneal dystrophy.

Mitochondrial Features in *TGFBI*-Related Corneal Dystrophy

Mitochondria are essential structures that are primarily responsible for the production of energy and play an important role in cell metabolism and survival (Vallabh et al. 2017). Aging causes an accumulation of harmful proteins and results in mitochondrial dysfunction (Lane et al. 2015). Dysfunction and mutation of mitochondrial DNA are linked to development of numerous diseases. Mitochondria not only have proteins that are related with nuclear DNA but also have genes that are independent of nuclear DNA (Vallabh et al. 2017). The human mitochondrial genome is in a circular form of 16,569 bp and contains 37 genes that maintain mitochondrial function. Among these genes, 13 genes encode the enzymes required for the oxidative phosphorylation pathway that governs ATP production (Lascaratos et al. 2012). Corneal fibroblasts in the anterior stroma have twice as many mitochondria as those in the posterior stroma, which correlates with the higher oxygen tension and cell density of the anterior stroma (Vallabh et al. 2017).

Corneal fibroblasts from patients with GCD2 exhibit swollen and degenerative mitochondria with an increased number in early passages and a decreased number of disorganized and dilated mitochondria with abrupt loss of mitochondrial activity and mitochondrial membrane potential in late passages (Fig. 8.4) (Kim et al. 2011). About

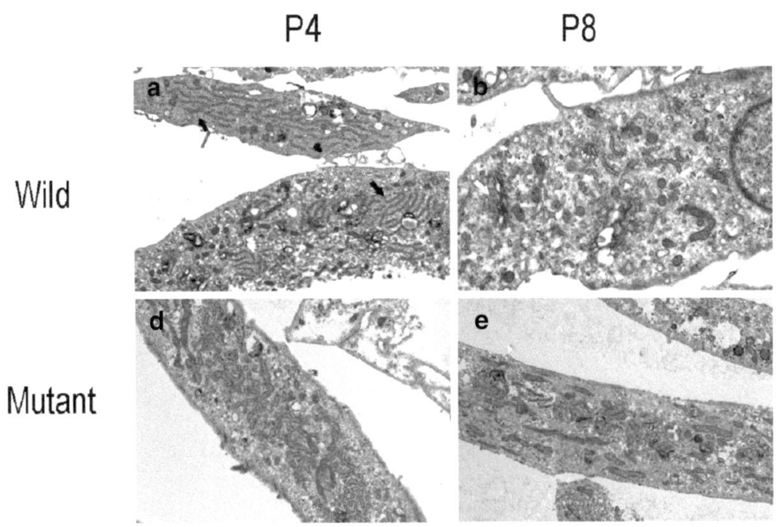

Fig. 8.4 Electron microscopic images of cultured corneal fibroblasts. Wild-type cultured fibroblasts contain small round mitochondria and other cellular components such as endoplasmic reticulum (black arrow) in passage 4 (**a**) and Golgi complex (white arrow) in passage 8 (**b**). (**c**) Homozygous GCD2 cultured fibroblasts demonstrate swollen mitochondria with increasing numbers throughout the cytoplasm and limited numbers of other intracellular organelles. (**d**) In late-passage homozygous GCD2 corneal fibroblasts, disorganized and dilated mitochondria with decreased numbers are observed. Reprinted from (Kim et al. 2011) with permission from Elsevier

23% of wild-type corneal fibroblasts contain fragmented mitochondria, whereas about 96% of GCD2 corneal fibroblasts contain fragmented mitochondria (Choi et al. 2012). Impaired proliferation, reduced viability, and increases in apoptotic features were identified in GCD2 corneal fibroblasts. The expression levels of mitochondrial complex I, II, and IV were significantly reduced in late passage GCD2 corneal fibroblasts. A *TGFBI* mutation may be associated with alteration of mitochondrial structure and function, which can cause cellular damage and delayed proliferation of aged cells. Treatment with an antioxidant, butylated hydroxyanisole, recovered increased ROS levels in GCD2 corneal fibroblasts, which may rescue *TGFBI* mutation-induced cellular toxicities in keratocytes.

Autophagy in *TGFBI*-Related Corneal Dystrophy

The balance between the synthesis and degradation of macromolecules and organelles is important for cellular homeostasis. The autophagy pathway and the ubiquitin-proteasome system are two main degradation pathways that govern the removal of dysfunctional proteins (Choi and Kim 2016). Many studies have reported that the accumulation of disease-related proteins is intimately involved in both autophagy and ubiquitin-proteasome pathways (Levine and Kroemer 2008; Maynard et al. 2009). The ubiquitin-proteasome system mainly degrades short-lived or soluble proteins after being tagged by ubiquitin to be recognized by proteasome under tight regulation (Ciechanover et al. 2000), and the autophagy pathway primarily breaks down aggregated or long-lived proteins and intracellular organelles (Mizushima et al. 2008). Recently, the interaction and cross-talk between autophagy and ubiquitin-proteasome systems has been extensively studied (Choi and Kim 2016).

The autophagy pathway degrades macromolecules and disorganized or excess organelles after encapsulation with autophagosome (Mizushima 2007). Subsequently, autophagosomes fuse with lysosomes, so that the bulky cellular macromolecules and organelles are degraded by lysosomal enzymes. The autophagy pathway can be classified into three subtypes: macroautophagy, microautophagy, and chaperone-mediated autophagy. Macroautophagy has been actively investigated and considered the principal pathway for protein and organelle degradation (Han et al. 2016). The mammalian target of rapamycin (mTOR) is one of the major regulators of autophagy (Choi and Kim 2016) (Fig. 8.5). mTOR signaling pathway negatively controls autophagy, and mTOR interacts with other regulatory molecules, including G-protein β-subunit-like protein (GβL) to form two main compounds: mTORC1 and mTORC2. mTORC1 has regulatory associated protein of mTOR (raptor) as a subunit, so it is sensitive to rapamycin, while mTORC2 has rapamycin-insensitive companion of mTOR (rictor), which makes it insensitive to rapamycin (Hara et al. 2002). The phosphatidylinositol 3-kinase/serine/threonine kinase AKT (PI3K/AKT) and mitogen-activated protein kinase/extracellular signal-regulated kinase 1/2 (MAPK/ERK 1/2) inhibit the autophagy pathway by activating the mTOR pathway as upstream regulators (Choi and Kim 2016). mTOR signaling is suppressed by p53, which can promote the initiation of the autophagy (Feng et al. 2005). The autophagy pathway is induced by Beclin1, which is an mTOR-independent activator of autophagy, and is suppressed by anti-apoptotic proteins, Bcl-2 and Bcl-XL (Pattingre et al. 2005). In addition, autophagy is regulated by stress and pathological conditions, including oxidative stress and endoplasmic reticulum stress (Choi and Kim 2016).

In corneal fibroblasts, the autophagy pathway works to help maintain cellular homeostasis. Oxidative stress has been reported to cause the accumulation of mitochondria by disrupting the autophagy pathway (Luo et al. 2013). Oxidative stress, changes in mitochondrial morphology and numbers, and accumulation of mutant TGFBIp in lysosomes, which are the pathologic characteristics found in GCD2 corneal fibroblasts, suggest that autophagy is involved and defective. The level of intracellular TGFBIp was significantly increased after treatment of cultured wild-type

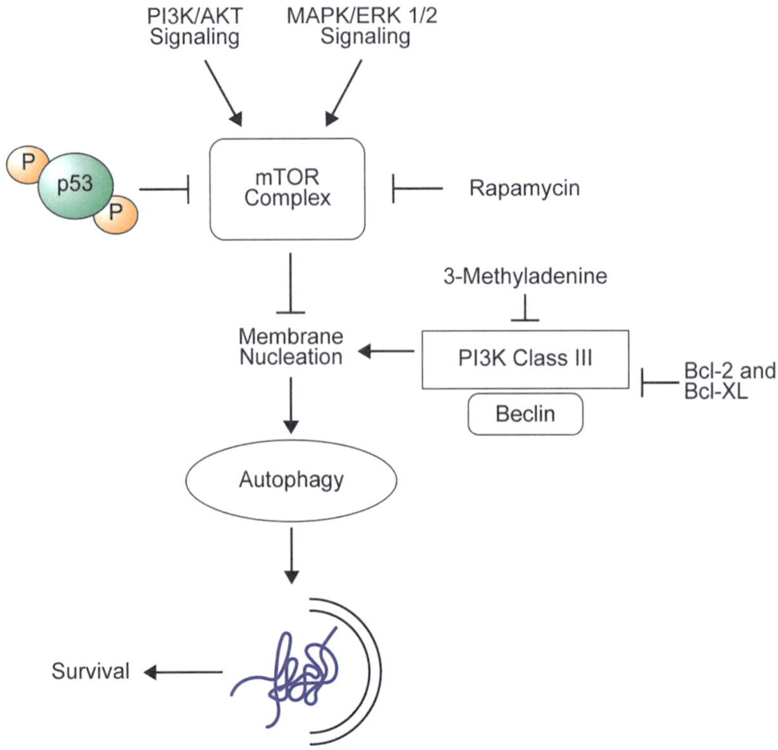

Fig. 8.5 Schematic diagram of regulation of autophagy pathway in mammalian cells. The phosphatidylinositol-3-kinase (PI3K)/the serine/threonine kinase AKT and mitogen-activated protein kinase (MAPK)/extracellular signal-regulated kinase 1/2 (ERK 1/2) signaling pathways activate mammalian target of rapamycin (mTOR) and sequentially block the activation of autophagy. p53 suppresses mTOR, thus facilitating autophagy, as a downstream target. A class III PI3K complex (Beclin, Class III PI3K, and p150) initiates membrane nucleation by recruiting microtubule-associated protein light chain 3 (LC3) and autophagy-related proteins, ATG. The phagophore is generated either from preexisting ER membrane or de novo. After autophagosome formation, it fuses with the lysosome to become an autophagolysosome, where lysosomal hydrolases digest the sequestered cytoplasmic-derived contents. Autophagy can be activated with AKT inhibitors and rapamycin, a small molecular inhibitor to mTOR, and be blocked with 3-methylamphetamine (3-MA), an inhibitor for class III PI3K

and GCD2 corneal fibroblasts with the autophagy inhibitor, bafilomycin A1, while treatment of MG132, a ubiquitin-proteasome system inhibitor, did not change the level of TGFBIp (Han et al. 2016). This finding indicates that the autophagy pathway is the major degradation pathway of TGFBIp, rather than ubiquitin-proteasome system. The expression of LC3 II, which is a marker for autophagy, was increased in GCD2 corneal fibroblasts, and wild-type corneal fibroblasts are genetically modified to overexpress mutant TGFBIp (Fig. 8.6) (Choi et al. 2012). Treatment of corneal fibroblasts from patients with GCD2 with leupeptin, an inhibitor of lysosomal cathepsins, causes accumulation of LC3 II and suppresses degradation rates of TGFBIp (Choi et al. 2014). The expression of raptor, a component of mTORC1, which inhibits autophagy pathway, was significantly reduced in GCD2 corneal fibroblasts. These results suggest that the autophagy pathway is activated in GCD2.

Mutant TGFBIp accumulates in autophagosome with LC3-enriched vesicles and lysosomes in GCD2 corneal fibroblasts. Although the amount of TGFBIp was not significantly different between GCD2 homozygous and wild-type corneal fibroblasts, the level of *TGFBI* mRNA was significantly lower in GCD2 corneal fibroblasts. The expression of LC3 and P62 was lower in wild type than GCD2, suggesting that blockage

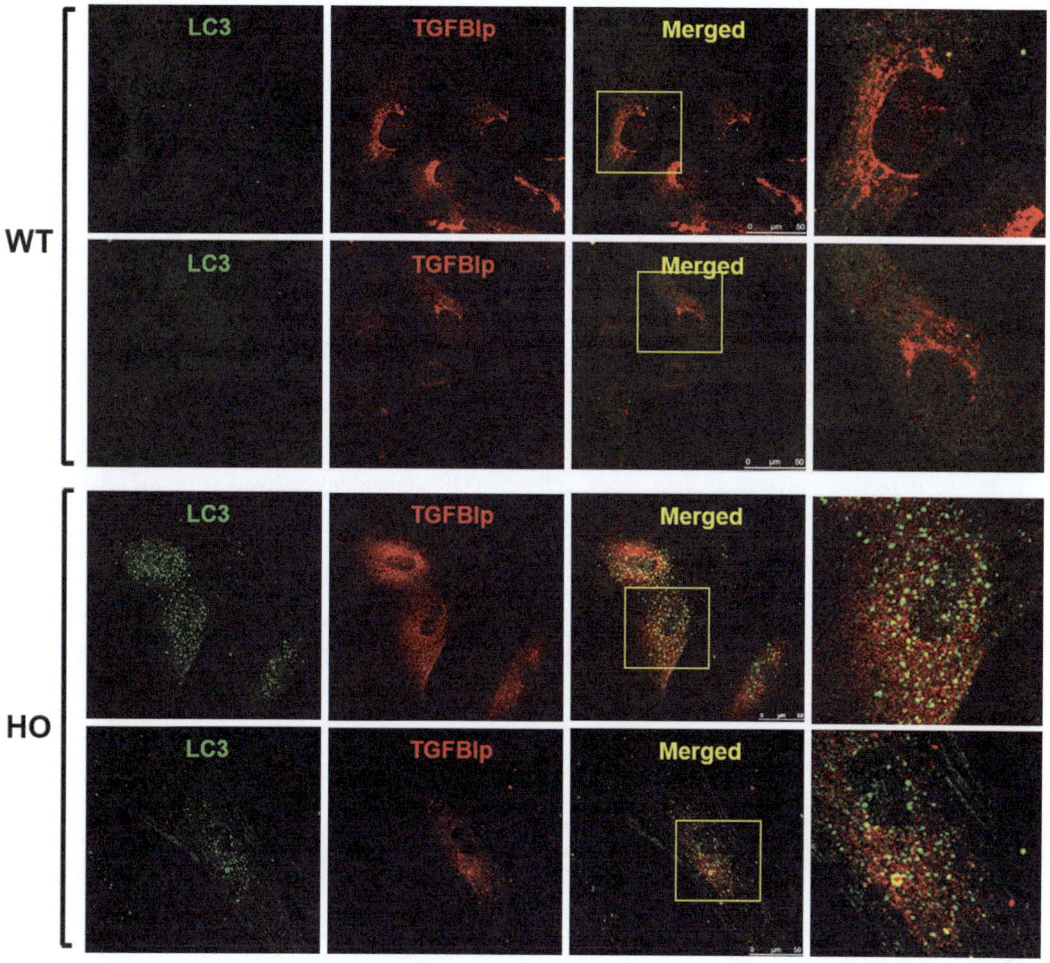

Fig. 8.6 Immunofluorescence images of endogenous LC3 and TGFBIp in wild-type (WT) and GCD2 homozygote (HO) corneal fibroblasts. LC3 showed vesicular staining, and the number of LC3-positive puncta was higher in HO corneal fibroblasts than in control cells. Insets show a 3.5-fold magnification of the indicated region. A merged image of the red and green channels is shown in the third picture in each row; yellow indicates overlapping localization. Reprinted from (Choi and Kim 2016) with permission from Elsevier

of fusion between autophagosome and lysosomes explains why the difference between wild-type and GCD2 corneal fibroblasts was small. When cycloheximide, a protein synthesis inhibitor, was treated in cultured corneal fibroblasts, TGFBIp was almost absent in wild type after 40 min, but it was still detected after 60 min in GCD2. Primary cultured wil-type corneal fibroblasts designed to express mutant TGFBIp also showed an accumulation of intracellular TGFBIp. These findings are thought to be the result of a damaged degradation pathway or delayed extracellular secretion. In addition, fragmented and disorganized mitochondria that require mitochondrial autophagy to be removed were present in GCD2 corneal fibroblasts (Choi and Kim 2016). Overall, although autophagy is activated in GCD2, there is a functional defect that is not sufficient to completely degrade the mutant TGFBIp, suggesting accumulation of mutant protein.

The autophagy pathway can play a role in protecting cells, but many studies have been shown that it also contributes to cell death (Denton and Kumar 2019; Bialik et al. 2018). Treatment with an autophagy inhibitor significantly reduced cellular proliferation of GCD2 corneal fibroblast more than it did for wild-type corneal fibroblasts. In addition, bafilomycin A1 (autophagosome-lysosome fusion inhibitor) induced caspase-3 and poly (ADP-ribose) polymerase 1 (PARP1) activation in GCD2 corneal fibroblasts (Choi et al. 2012). These results suggest that activation of autophagy may play a protective role, and defective autophagy can lead to dysfunction of corneal fibroblasts in GCD2. Also, rapidly dividing cells may help reduce the toxicity related to the accumulation of misfolded or aggregated protein, as cell division would dilute the misfolded or aggregated protein in each cell. Meanwhile, corneal fibroblasts may be more susceptible to cytotoxicity due to defective autophagy than other cell types, because corneal fibroblasts are stationary, remain in the resting phase, and have a longer lifespan (Francesconi et al. 2000).

Because TGFBIp accumulates in autophagolysosomes in GCD2 corneal fibroblasts, regulation of autophagy to reduce the accumulation of mutant TGFBIp can be a potential therapeutic option for the treatment of *TGFBI*-related corneal dystrophy (Choi and Kim 2016). Rapamycin reduced expression of mutant TGFBIp in GCD2 corneal fibroblasts, while it did not change the level of wild-type TGFBIp, which indicates that rapamycin and related molecules may be effective for treatment of *TGFBI*-related corneal dystrophies. Rapamycin has the advantage that it is currently clinically available and that it only affects cells with defective autophagy. Melatonin has been shown to activate autophagy in wild-type and GCD2 corneal fibroblasts through an mTOR-dependent pathway (Fig. 8.7) (Choi et al. 2013). Treatment with melatonin increases Beclin 1 which is associated with formation and maturation of autophagosome and reduces the autophagy-inhibitory effects of bafilomycin A1. Melatonin significantly increases TGFBIp degra-

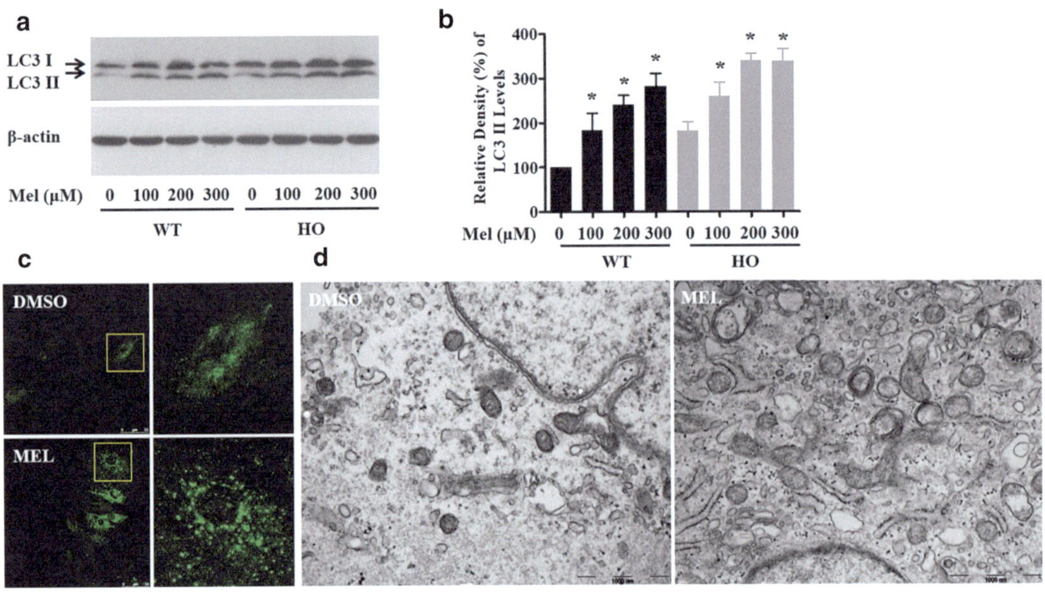

Fig. 8.7 Autophagosome formation was facilitated by melatonin in wild-type (WT) and GCD2 homozygote (HO) corneal fibroblasts. (**a, b**) LC3 conversion was analyzed 12 h. After melatonin treatment in corneal fibroblasts by Western blot assay. Melatonin induced significantly the level of LC3 II in both WT and GCD2 HO corneal fibroblast. (**c**) Autophagosome formation was analyzed by fluorescence microscopy. Corneal fibroblasts were cultured for 12 h. in the absence (upper panel) or presence (lower panel) of 100 μM melatonin, which showed increased autophagosome formation with melatonin. (**d**) Transmission electron micrographs of corneal fibroblasts were treated with 100 μM melatonin or vehicle alone (DMSO). *AV* autophagosome, *N* nuclear, *m* mitochondria. John Wiley & Sons Ltd. for permission to reprint Choi SI, et al. 'Melatonin induces autophagy via an mTOR-dependent pathway and enhances clearance of mutant TGFBIp', *Journal of Pineal Research* 54(2013):361–372

dation and reduces the expression of wild-type and mutant TGFBIp. Furthermore, simultaneous treatment with both melatonin and rapamycin enhances the clearance of mutant TGFBIp when compared to the clearance under treatment with either drug alone. Taken together, melatonin might be a potential therapeutic strategy for *TGFBI*-related corneal dystrophies. Excessive or sustained activity of autophagy can lead to apoptosis cell death, so autophagy can function as a double-edged sword (Choi and Kim 2016). Thus, more research is needed to discover the possibility of autophagy regulation as a therapeutic option for the treatment of *TGFBI*-related corneal dystrophies.

Recently, a study has been reported that reduced autophagolysosomal degradation in GCD2 is associated with a decrease in cathepsin and an increase in lysosomal pH. Cathepsins K and L were significantly reduced in GCD2 corneal fibroblasts, but cathepsins B and D did not show differences (Choi et al. 2020). Treatment with cathepsin L inhibitor significantly increased the accumulation of TGFBIp in both wild-type and GCD2 cells, and exogenous cathepsin L expression promoted the clearance of mutant TGFBIp, but not of normal TGFBIp in wild-type corneal fibroblasts. The expression of transcription factor EB (TFEB), which functions as a master regulator of lysosomal biogenesis, was significantly decreased in GCD2 corneal fibroblasts, and exogenous TFEB expression rescued lysosomal function and mutant-TGFBIp clearance in GCD2 cells.

Endocytosis of TGFBIp

Secretory proteins usually contain signal peptides, which sort the protein well to the ER. These proteins are mainly transferred to the ECM or the plasma membrane through the ER/Golgi secretory pathway, although an unconventional secretory pathway (non-ER/Golgi secretory pathway) is used for trafficking of some proteins in certain environments (Choi et al. 2015b). After the proteins move to the ER, proteins are packaged into vesicles that are transferred to endosomal or lysosomal compartments, or to the plasma membrane directly. As an alternative, the protein may be packaged in uncoated vesicles that are targeted to the Golgi apparatus before reaching the ECM, or fuse directly the plasma membrane. The autophagy pathway delivers cytosolic proteins and organelles to lysosomes for degradation and regulates unconventional protein secretion, as well.

Endocytosis is one of the basic cellular mechanisms of eukaryotic cells that induces internalization of molecules from the surface of cells or ECM. The proteins internalized from the plasma membrane are recycled to the cellular membrane or transported to lysosomes for degradation. Endocytosis includes phagocytosis, which internalizes large molecules, and pinocytosis, which internalizes fluids and solutes (Choi et al. 2015b). Endocytosis has an important role in regulating ECM turnover by removal of dysfunctional molecules in lysosomal trafficking. Degradation and elimination of ECM proteins are strongly associated with tissue development, remodeling, and repair (Choi et al. 2015b). ECM proteins could also be removed by matrix metalloproteases and other proteases extracellularly. Many diseases such as fibrosis, arthritis, and cancer can be caused by impairment of ECM homeostasis.

TGFBIp is mainly secreted through the conventional ER-Golgi secretory pathway (Choi et al. 2015b). Treatment with brefeldin-A, which inhibits transport from the ER, almost completely suppressed secretion of TGFBIp by both wild-type and GCD2 corneal fibroblasts. Inhibition of protein transport from medial to trans-Golgi cisternae with monensin also nearly completely blocked the extracellular secretion of TGFBIp. Dynamic monitoring of the secretion of TGFBIp using pulse-chase experiments has shown that newly synthesized TGFBIp in wild type is almost totally secreted into the culture media with only 0.5% remaining intracellularly, while, in GCD2 cells, 20% of TGFBIp is retained inside the corneal fibroblasts. Notably, TGFBIp secretion is delayed in GCD2 corneal fibroblasts compared to that in normal corneal fibroblasts (Choi et al. 2015b).

Internalization of TGFBIp was confirmed in several cell types including NIH3T3, ZW13–1, and SK-N-MC cell lines which do not express endogenous TGFBIp (Choi et al. 2015b). Treatment with genistein, which blocks caveolae-mediated endocytosis in corneal fibroblasts, decreased the amount of intracellular TGFBIp, whereas chlorpromazine, an inhibitor of clathrin-mediated internalization, did not suppress the intracellular TGFBIp expression. Nystatin treatment, which inhibits caveolae-mediated internalization, also reduced the level of intracellular TGFBIp. In addition, a caveolin-1 knockout cell line did not show internalization of TGFBIp. These findings suggest that caveolae-mediated endocytosis is the main internalization pathway for TGFBIp. Caveolin-1 co-localized with TGFBIp and with cathepsin D in corneal fibroblasts. Inhibition of lysosomal degradation increased internalized TGFBIp that had accumulated in lysosomes. Flask-shaped invagination of the cell membrane and abundant membrane-associated vesicles were observed in corneal fibroblasts with transmission electron microscopy, and the caveolae were clustered into caveosomes. Immunogold-labeled TGFBIp was detected in caveolae on the plasma membrane. Furthermore, both caveolae-like vesicles and labeled TGFBIp were located in lysosomes. Caveolae-mediated internalized TGFBIp is transported to lysosomes and degraded.

Integrins are known to be important molecules for endocytosis and recycling. TGFBIp has also been reported to interact with integrins including $\alpha_V\beta_3$ through RGD binding motifs (Han et al. 2016). Incubation of exogenous TGFBIp with RGD peptide significantly decreases the amount of internalized TGFBIp. TGFBI was bound to integrin α_V or $\alpha_V\beta_3$ in immunoprecipitation experiments, and co-localization of integrin α_V and internalized TGFBIp was localized by confocal microscopy to corneal fibroblasts. Taken together, these findings indicate that TGFBIp is internalized through the interaction of the RGD-motif in TGFBIp with integrins (Choi et al. 2015b). A schematic diagram of secretion and internalization of TGFBIp is shown in Fig. 8.8.

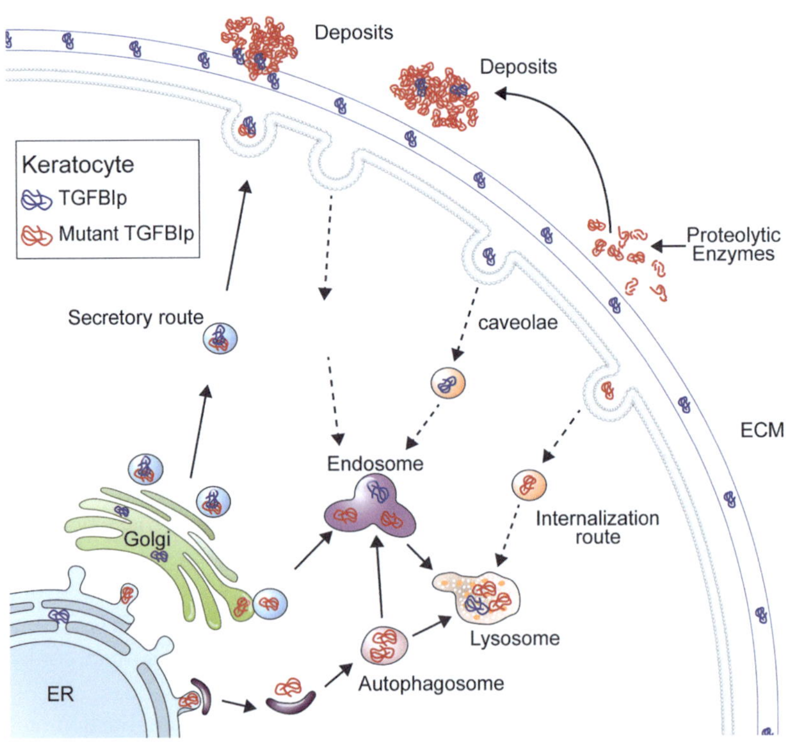

Fig. 8.8 Schematic diagram of TGFBIp trafficking. TGFBIp is mainly secreted through the ER-Golgi system. Secreted TGFBIp plays a role in the ECM and undergoes remodeling and turnover. Secreted mutant TGFBIp aggregates with other ECM molecules, such as collagen. Mutated TGFBIp is transported to and destroyed by the lysosome via the autophagy pathway. A certain portion of secreted TGFBIp can be internalized through caveolae-mediated endocytosis and transported to lysosomes for degradation

Altered Cell Cycle in GCD2 Corneal Fibroblasts

The cell cycle of eukaryotic cells can be divided in two main periods: the interphase and mitosis. The interphase consists of the G1 phase, in which cells decide to grow or enter a quiescent phase G0, S phase where DNA synthesis occurs, and G2 phase where cells are preparing for mitosis. The second phase, M phase, during which cells divide the genetic material and cytokinesis occurs, has a shorter period (Leal-Esteban and Fajas 2020). The transition from phase to phase in the mammalian cell cycle is tightly regulated by the cyclins, cyclin-dependent kinases (CDKs), CDK inhibitors (CKI), and the retinoblastoma family of proteins (Norbury and Nurse 1992). Cyclins, CKIs, and tumor-suppressor protein, p53 are controlled by ubiquitin-mediated proteolysis (Choi et al. 2015a).

GCD2 corneal fibroblasts showed morphological variations including an increase in size and decreased proliferation rate compared to wild type (Kim et al. 2011). In cell cycle analysis using FACS, wild-type corneal fibroblasts accumulate in the G1 phase; however, GCD2 corneal fibroblasts were arrested G2/M phase. After serum starvation in wild-type and GCD2 corneal fibroblasts, which caused cell cycle synchronization, the proportion of cells in G1, S, and G2/M phase was comparable (Choi et al. 2015a). Incubation for 48 h with serum after synchronization of cell cycle using serum starvation showed altered cell cycle proportion between wild-type and GCD2 corneal fibroblasts. S phase cells were significantly increased in GCD2 cells, compared to wild type cells, while G1 phase and G2/M phase cells were significantly decreased. These findings suggest that GCD2 corneal fibroblasts have a disrupted cell cycle.

The expression of Cyclin A1, Cyclin B1, Cyclin D1, Cyclin E1, CDK6, p21, and p53 was significantly decreased, whereas p16 and p27 expression was higher in GCD2 corneal fibroblasts than wild-type cells (Choi et al. 2015a). After serum starvation, the expression of Cyclins A1 and B1 was decreased in wild-type cells, reaching similar levels of GCD2 cells. Re-introduction of serum after starvation showed that the expression of Cyclins A1 and B1 was increasing; however, GCD2 corneal fibroblasts were observed to be delayed in expression compared to control cells. The expression of Cyclins A1, B1, and D1 and p53 was significantly reduced, and the expression of p21 was induced after treatment of bafilomycin A1 in the wild-type corneal fibroblasts. Inhibition of the ubiquitin-proteasome system with MG132 increased the expression of those proteins affected by bafilomycin A1, which indicates that cell cycle-related proteins are degraded by the ubiquitin-proteasome system, when the autophagy pathway is damaged (Choi et al. 2015a). These findings indicate that cell cycle arrest and dysfunction which are associated with defective autophagy reduce cellular proliferation in *TGFBI*-related corneal dystrophies.

Regulation of TGFBIp Expression in *TGFBI*-Related Corneal Dystrophy

TGFBIp, as the name suggests, is a protein whose expression is increased by TGF-β. Also, in corneal epithelium and stroma, TGF-β1 induces *TGFBI* mRNA expression and extracellular TGFBIp accumulation (Yellore et al. 2011; Jeon et al. 2012). TGF-β1 regulates the amount of TGFBIp via mechanisms that depend on Smad transcription factors and E-box (Han et al. 2016). When TGF-β1 is bound to the TGFB1 receptor, activated receptor phosphorylates Smad2 and Smad3 and then Smad4 assemble together. The Smad complex then moves into the nucleus and controls expression of target genes (Shi and Massague 2003). In corneal fibroblasts, Smad3 has been demonstrated to play a key role in TGF-β1-associated TGFBIp expression (Jeon et al. 2012).

Lysophosphatidic acid (LPA) is generated by activated adipocytes and platelets, fibroblasts, and cancer cells during inflammation and wound healing. It has important roles in cell survival, proliferation, migration, and adhesion. LPA_1 and LPA_3 receptors are expressed in normal intact corneas, and LPA concentrations are increased after corneal injury (Wang et al. 2002). LPA facilitates proliferation of corneal epithelial cells, keratocytes, and endothelial cells and regulates corneal homeostasis and wound healing (Jeon et al. 2012). LPA induced TGFBIp expression via augmentation of TGF-β1 secretion in wild-type and GCD2 corneal fibroblasts, which was inhibited by LPA receptor antagonists and knockdown of LPA_1 receptor. Neutralizing antibody for TGF-β1 and an inhibitor of TGF-β type I receptor kinase decreased LPA-induced TGFBIp expression. Smad2, Smad3, and JNK have been found to play key roles in the regulation of TGFBIp related with LPA (Fig. 8.9) (Jeon et al. 2012).

TGFBIp expression is also regulated by chromatin conformation at the *TGFBI* promoters (Lee et al. 2011). The promoter region in TGFBIp expressing cells showed higher accessibility to restriction enzyme than in cells not expressing TGFBIp, which indicates that open chromatin conformation is advantageous for TGFBIp expression. Transcription factors, Sp1 and Sp3, bind *TGFBI* promoter and are required for TGFBIp expression. In addition, mutated SP1 elements, Sp1 inhibitor, and knockdown of Sp1 diminished *TGFBI* promoter activity, while overexpression of Sp1, or Sp3 augmented it (Lee et al. 2011). Factors Sp1 and Sp3 are associated with the regulation of constitutive TGFBIp expression.

Screening of potential microRNA (miRNA) targets for *TGFBI* showed the miR-9, miR-21, and miR-181a to be possible targets (Choi et al. 2016b). TGFBIp expression was significantly reduced by miR-9, miR-21, and miR-181a in corneal fibroblasts and was significantly enhanced

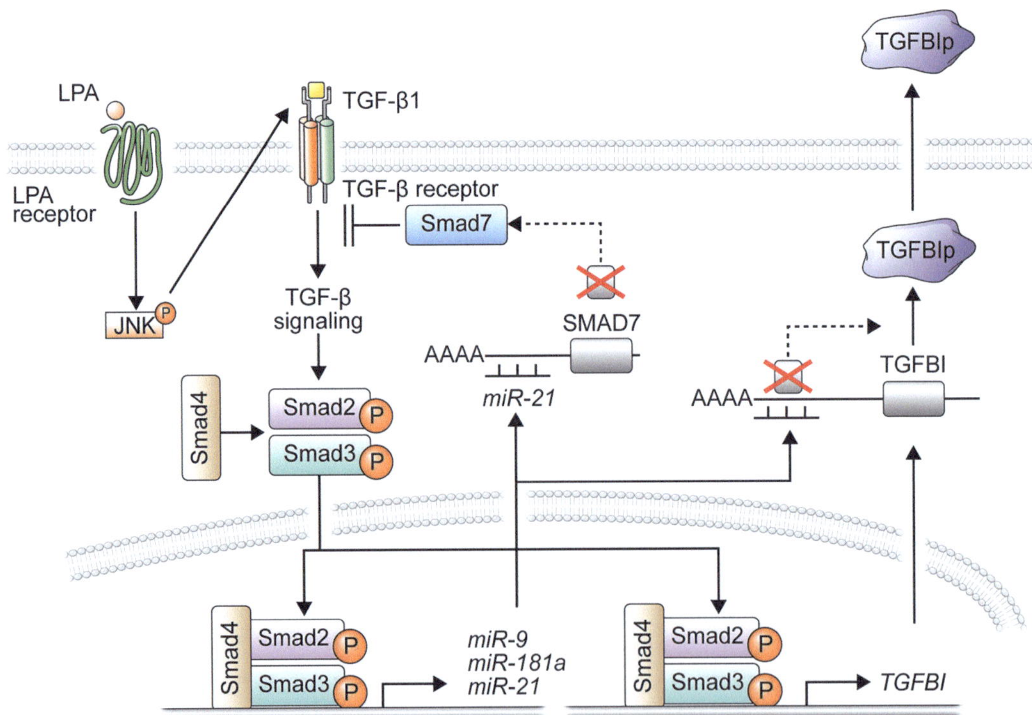

Fig. 8.9 Schematic illustration of cellular signaling and molecular mechanism of TGFBIp expression. TGF-β binds the TGF-β receptor and activates TGF-β signaling, including Smad2/3 phosphorylation. TGF-β induces *TGFBI* mRNA expression, as well as the expression of miR-9, miR-181a, and miR-21 through Smad2/3 signaling. Therefore, TGF-β provides dual function in regulation of TGFBIp. TGF-β augments TGFBIp expression via upregulation of *TGFBI* mRNA expression. Conversely, TGF-β downregulates TGFBIp expression, since miR-9, miR-181a, and miR21 decrease *TGFBI* mRNA. Additionally, miR-21 reduces Smad7, which acts as a TGF-β signaling inhibitor. Lysophosphatidic acid (LPA) activates JNK signaling pathway via LPA receptor, which activates TGF-β signaling pathway

by anti-miR-21. The amount of miR-21 was significantly higher than those of miR-9 and miR-181a in corneal fibroblasts. Interestingly, TGF-β1 increased both TGFBIp and levels of miR-21 and miR-181a, indicating ECM homeostasis is mediated by the coordination of TGF-β and miRNAs (Fig. 8.9) (Choi et al. 2016b).

Epigenetic regulation is also involved in TGFBIp expression. Posttranslational modification of nucleosome histones, such as acetylation, methylation, and ubiquitination, plays an important role in regulating chromatin structure and transcription (Maeng et al. 2015b). Generally, the acetylation of H3 histone lysines (H3KAc), which is mediated by histone acetyl transferases, increases transcription of genes, while the methylation of H3 lysines (H3Kme) mediated by histone methyltransferase (HMTs) activates or inactivates transcription of genes depending on the location of lysine (Maeng et al. 2015b). Methylation of the fourth lysine residue at H3 (H3K4me) is usually related to transcription elongation and amplification of genes. In contrast, methylation of the ninth lysine at H3 (H3K9me) usually represses gene expression (Martin and Zhang 2005). H3K4me3 in Smad Binding Element (SBE) of the *TGFBI* promoter was significantly lower in GCD2 corneal fibroblasts, compared to normal fibroblasts, as TGFBIp expression was lower in GCD2 (Maeng et al. 2015b). Occupancy of H3K4 methyltransferase, MLL1 and SET7/9, was increased in *TGFBI* promoter by TGF-β1 in wild-type corneal fibroblasts, whereas this effect was weak in GCD2 cells. Knockdown of MLL1 and SET7/9 decreased H3K4me levels and expression of TGFBIp induced by TGF-β1 (Maeng et al. 2015b).

Pathogenic Mechanism of Corneal Deposits in *TGFBI*-Related Corneal Dystrophy

Under normal conditions, TGFBIp is produced primarily by the corneal epithelium; however, TGFBIp is expressed in both the corneal epithelium and keratocytes in cases of corneal damage (Takacs et al. 1999). In Reis-Bücklers corneal dystrophy and Thiel-Behnke corneal dystrophy, mutated TGFBIp aggregates primarily in the Bowman layer. In GCD1, the deposits are found in the superficial stroma, as well as in epithelium, and Bowman layer (Akhtar et al. 1999), while most of deposits are located in the superficial stroma (hyaline) and deep stroma (amyloid) in GCD2 (Han et al. 2016). In lattice corneal dystrophy (LCD), deposits are shown as the central superficial haze and paracentral stromal lattice lines (Lisch and Seitz 2014). When observing recurrent LCD1 cornea after keratoplasty, only the diffuse and superficial opacities were noted, and stromal lattice lines were not observed. Lisch and Seitz suggest that the recurrence of LCD1 after keratoplasty is mainly caused by TGFBIp, which is produced by the corneal epithelium, and the deep stromal lattice line is thought by some to be caused by TGFBIp produced by keratocytes (Lisch and Seitz 2014). In contrast, some researchers believe that the corneal deposits originate from corneal epithelium, because recurrent granular deposits are primarily located in the subepithelial layer after corneal transplantation in GCD (Frising et al. 2006). More details about the components of lattice corneal dystrophy will be discussed in Chap. 9.

Notably, mutations in *TGFBI* are known to cause only dystrophies in the cornea. Pathologic deposits caused by TGFBIp accumulation were only found in the cornea, and not in other tissues or organs, including the brain, heart, lung, liver, kidney, spleen, esophagus, bone marrow, lymph nodes, aorta, and bladder in autopsy of a patient with LCD1 (El Kochairi et al. 2006). This may be related with the high concentration of TGFBIp in the cornea. In addition, the specific factors in the cornea may induce aggregation of mutated TGFBIp, which is supported by research showing that LCD1 corneal fibroblasts do not spontaneously generate amyloid (Korvatska et al. 2000), and heparin sulfate promotes fibril formation in p.(Ala546Thr) mutation (Andreasen et al. 2012). The mouse cornea shows a different appearance even though the LCD1 p.(Arg124Cys) mutation is induced, and this fact suggests that components of the human cornea such as extracellular matrix or collagen types are responsible for the deposit formation in the human cornea (Kitamoto et al. 2020).

Clear zones are usually found between the deposits and limbus, and clear zones are also observed around the advancing edge of pterygia (Lee et al. 2006). In addition, damage at the center of the cornea, such as LASEK, often stimulates accumulation of corneal deposits, while no accumulation of deposit is noted near perilimbal cataract incisions (Han et al. 2016). These observations support the hypothesis that blood vessels can suppress the accumulation of mutant TGFBIp, and abnormal TGFBIp only affects the cornea because of the absence of the blood or blood vessels in the cornea (Lee et al. 2006).

Patients with the same *TGFBI* mutation may display different phenotypes, even within the same family. This is probably due to the fact that the corneal deposits are composed of not only abnormal TGFBIp, but also other ECM substances, and the ECM environment is different for each patient. One study, showing that the expression of 555 genes was different by more than two folds between wild-type and GCD2 corneal fibroblasts, supports this hypothesis (Choi et al. 2010). In addition, two mutations in *TGFBI* in an individual can occur and affect the phenotype (Han et al. 2016). The p.[(Ala546Asp;Pro551Gln)] in the same allele (cis) induced bilateral, symmetric, branching refractile lines in the central anterior stroma and multiple polymorphic refractile deposits in the mid and posterior stroma (Aldave et al. 2004). The p.[(Arg124Leu;Thr125_Glu126del)] cis mutation showed round or snowflakes-like deposits in the subepithelial layer and rapid spreading of the deposits (Dighiero et al. 2000). The p.[Arg124His];[Asn544Ser] mutations in different alleles (trans) showed both phenotypes of GCD2 and LCD (Yamada et al. 2009), and the p.[Arg549Thr];[Arg555Trp] trans compound heterozygote showed unusual severe phenotypes and profound recurrence after keratoplasty (Frising et al. 2006).

Abnormal degradation of mutated proteins is associated with pathogenesis of disease (Han et al. 2016). Changes in the proteolytic process and protein turnover rates were reported to be involved in the development of *TGFBI*-related corneal dystrophy. There is a difference in the characterization of protein turnover and degradation depending on the mutation type of *TGFBI* (Korvatska et al. 2000). Normal cornea contains major 68 kDa TGFBIp and minor species of 64, 57, 47, and 29 kDa fragments. The corneal deposits of p.(Arg124Cys) showed an amyloid pattern and accumulation of N-terminal TGFBIp, in which 44 kDa species were the major components of amyloid fibrils. The p.(Arg124Leu) cornea showed non-amyloid pattern with 68 kDa full-size form, and the p.(Arg124His) cornea showed mixed pattern of deposition with 68 and 66 kDa species (Korvatska et al. 2000). In HEK293FT cells transfected with LCD-associated mutant *TGFBI*, 47-kDa-fragmented TGFBIp was detected in p.(Leu518Pro), and 43-kDa-fragmented TGFBIp was detected in p.(Arg124Cys) (Han et al. 2011). Because these fragments are absent in wild type, they may have a potential role in corneal deposits in LCD. The synthetic TGFBIp 110–131 amino acid peptide was used for examining amyloidogenicity of the mutant, where p.(Arg124Cys) peptide showed higher Thioflavin T fluorescence than wild-type peptide, implying a higher capacity for amyloidogenicity of p.(Arg124Cys) (Schmitt-Bernard et al. 2000). Blockage of the SH group to inhibit disulfide bonds formation and deletion of 110–113 residue including Val-Val sequence inhibited formation of amyloid fibrils, suggesting cysteine-disulfide bonds and Val-Val hydrophobic interaction contributed to amyloidogenesis (Schmitt-Bernard et al. 2002). The proteolytic processing of TGFBIp in *TGFBI*-related corneal dystrophy will be covered in more detail in Chap. 9.

References

Akhtar S, et al. Deposits and proteoglycan changes in primary and recurrent granular dystrophy of the cornea. Arch Ophthalmol. 1999;117(3):310–21.

Aldave AJ, et al. Lattice corneal dystrophy associated with the Ala546Asp and Pro551Gln missense changes in the TGFBI gene. Am J Ophthalmol. 2004;138(5):772–81.

Andreasen M, et al. Polymorphic fibrillation of the destabilized fourth fasciclin-1 domain mutant A546T of the Transforming growth factor-beta-induced protein (TGFBIp) occurs through multiple pathways with different oligomeric intermediates. J Biol Chem. 2012;287(41):34730–42.

Bialik, S., S.K. Dasari, and A. Kimchi, Autophagy-dependent cell death—where, how and why a cell eats itself to death. J Cell Sci, 2018. 131(18) jcs215152.

Carson DD, et al. Changes in gene expression during the early to mid-luteal (receptive phase) transition in human endometrium detected by high-density microarray screening. Mol Hum Reprod. 2002;8(9):871–9.

Choi SI, Kim EK. Autophagy in granular corneal dystrophy type 2. Exp Eye Res. 2016;144:14–21.

Choi SI, et al. Decreased catalase expression and increased susceptibility to oxidative stress in primary cultured corneal fibroblasts from patients with granular corneal dystrophy type II. Am J Pathol. 2009;175(1):248–61.

Choi SI, et al. Involvement of TGF-{beta} receptor- and integrin-mediated signaling pathways in the pathogenesis of granular corneal dystrophy II. Invest Ophthalmol Vis Sci. 2010;51(4):1832–47.

Choi SI, et al. Impaired autophagy and delayed autophagic clearance of transforming growth factor beta-induced protein (TGFBI) in granular corneal dystrophy type 2. Autophagy. 2012;8(12):1782–97.

Choi SI, et al. Melatonin induces autophagy via an mTOR-dependent pathway and enhances clearance of mutant-TGFBIp. J Pineal Res. 2013;54(4):361–72.

Choi SI, et al. Autophagy is induced by raptor degradation via the ubiquitin/proteasome system in granular corneal dystrophy type 2. Biochem Biophys Res Commun. 2014;450(4):1505–11.

Choi SI, et al. Disrupted cell cycle arrest and reduced proliferation in corneal fibroblasts from GCD2 patients: a potential role for altered autophagy flux. Biochem Biophys Res Commun. 2015a;456(1):288–93.

Choi SI, et al. Lysosomal trafficking of TGFBIp via caveolae-mediated endocytosis. PLoS One. 2015b;10(4):e0119561.

Choi SI, et al. 4-Phenylbutyric acid reduces mutant-TGFBIp levels and ER stress through activation of ERAD pathway in corneal fibroblasts of granular corneal dystrophy type 2. Biochem Biophys Res Commun. 2016a;477(4):841–6.

Choi SI, et al. TGF-beta regulates TGFBIp expression in corneal fibroblasts via miR-21, miR-181a, and Smad signaling. Biochem Biophys Res Commun. 2016b;472(1):150–5.

Choi SI, et al. Melatonin reduces endoplasmic reticulum stress and corneal dystrophy-associated TGFBIp through activation of endoplasmic reticulum-associated protein degradation. J Pineal Res. 2017;63(3):e12426.

Choi SI, Woo JH, Kim EK. Lysosomal dysfunction of corneal fibroblasts underlies the pathogenesis of Granular Corneal Dystrophy Type 2 and can be rescued by TFEB. J Cell Mol Med. 2020;24(18):10343–55.

Ciechanover A, Orian A, Schwartz AL. Ubiquitin-mediated proteolysis: biological regulation via destruction. Bioessays. 2000;22(5):442–51.

Denton D, Kumar S. Autophagy-dependent cell death. Cell Death Differ. 2019;26(4):605–16.

Dighiero P, et al. A novel variant of granular corneal dystrophy caused by association of 2 mutations in the TGFBI gene-R124L and DeltaT125-DeltaE126. Arch Ophthalmol. 2000;118(6):814–8.

Doliana R, et al. EMI, a novel cysteine-rich domain of EMILINs and other extracellular proteins, interacts with the gC1q domains and participates in multimerization. FEBS Lett. 2000;484(2):164–8.

Dyrlund TF, et al. Human cornea proteome: identification and quantitation of the proteins of the three main layers including epithelium, stroma, and endothelium. J Proteome Res. 2012;11(8):4231–9.

El Kochairi I, et al. Systemic investigation of keratoepithelin deposits in TGFBI/BIGH3-related corneal dystrophy. Mol Vis. 2006;12:461–6.

Escribano J, et al. cDNA from human ocular ciliary epithelium homologous to beta ig-h3 is preferentially expressed as an extracellular protein in the corneal epithelium. J Cell Physiol. 1994;160(3):511–21.

Fan Y, et al. TGF-beta-induced upregulation of malat1 promotes bladder cancer metastasis by associating with suz12. Clin Cancer Res. 2014;20(6):1531–41.

Feng Z, et al. The coordinate regulation of the p53 and mTOR pathways in cells. Proc Natl Acad Sci U S A. 2005;102(23):8204–9.

Ferguson JW, et al. The extracellular matrix protein betaIG-H3 is expressed at myotendinous junctions and supports muscle cell adhesion. Cell Tissue Res. 2003a;313(1):93–105.

Ferguson JW, et al. Developmental expression patterns of Beta-ig (betaIG-H3) and its function as a cell adhesion protein. Mech Dev. 2003b;120(8):851–64.

Francesconi CM, et al. Expression patterns of retinoblastoma and E2F family proteins during corneal development. Invest Ophthalmol Vis Sci. 2000;41(5):1054–62.

Frising M, et al. Recurrent granular dystrophy of the cornea: an unusual case. Cornea. 2006;25(5):614–7.

Gibson MA, Kumaratilake JS, Cleary EG. Immunohistochemical and ultrastructural localization of MP78/70 (betaig-h3) in extracellular matrix of developing and mature bovine tissues. J Histochem Cytochem. 1997;45(12):1683–96.

Hahnel C, et al. The keratocyte network of human cornea: a three-dimensional study using confocal laser scanning fluorescence microscopy. Cornea. 2000;19(2):185–93.

Han YP, et al. Unique TGFBI protein in lattice corneal dystrophy. Invest Ophthalmol Vis Sci. 2011;52(11):8401–6.

Han KE, et al. Pathogenesis and treatments of TGFBI corneal dystrophies. Prog Retin Eye Res. 2016;50:67–88.

Hara K, et al. Raptor, a Binding Partner of Target of Rapamycin (TOR). Mediates TOR Action. Cell. 2002;110(2):177–89.

Hashimoto K, et al. Characterization of a cartilage-derived 66-kDa protein (RGD-CAP/beta ig-h3) that binds to collagen. Biochim Biophys Acta. 1997;1355(3):303–14.

Hirano K, et al. Beta ig-h3 is synthesized by corneal epithelium and perhaps endotheliumin Fuchs' dystrophic corneas. Curr Eye Res. 1996;15(9):965–72.

Jeon ES, et al. Lysophosphatidic acid activates TGFBIp expression in human corneal fibroblasts through a TGF-beta1-dependent pathway. Cell Signal. 2012;24(6):1241–50.

Jester JV, et al. The cellular basis of corneal transparency: evidence for 'corneal crystallins'. J Cell Sci. 1999;112(Pt 5):613–22.

Kim HJ, Kim IS. Transforming growth factor-beta-induced gene product, as a novel ligand of integrin alphaMbeta2, promotes monocytes adhesion, migration and chemotaxis. Int J Biochem Cell Biol. 2008;40(5):991–1004.

Kim JE, et al. Identification of motifs in the fasciclin domains of the transforming growth factor-beta-induced matrix protein betaig-h3 that interact with the alphavbeta5 integrin. J Biol Chem. 2002;277(48):46159–65.

Kim MO, et al. Transforming growth factor-beta-inducible gene-h3 (beta(ig)-h3) promotes cell adhesion of human astrocytoma cells in vitro: implication of alpha6beta4 integrin. Neurosci Lett. 2003;336(2):93–6.

Kim TI, et al. Altered mitochondrial function in type 2 granular corneal dystrophy. Am J Pathol. 2011;179(2):684–92.

Kim EK, Lee H, Choi SI. Molecular pathogenesis of corneal dystrophies: schnyder dystrophy and granular corneal dystrophy type 2. Prog Mol Biol Transl Sci. 2015;134:99–115.

Kitahama S, et al. Expression of fibrillins and other microfibril-associated proteins in human bone and osteoblast-like cells. Bone. 2000;27(1):61–7.

Kitamoto K, et al. Generation of mouse model of TGFBI-R124C corneal dystrophy using CRISPR/Cas9-mediated homology-directed repair. Sci Rep. 2020;10(1):2000.

Klintworth GK, Enghild JJ, Valnickova Z. Discovery of a novel protein (big-h3) in normal human cornea. Invest Ophthalmol Vis Sci. 1994;35:1938.

Klintworth GK, Valnickova Z, Enghild JJ. Accumulation of beta ig-h3 gene product in corneas with granular dystrophy. Am J Pathol. 1998;152(3):743–8.

Korvatska E, et al. Amyloid and non-amyloid forms of 5q31-linked corneal dystrophy resulting from kerato-epithelin mutations at Arg-124 are associated with abnormal turnover of the protein. J Biol Chem. 2000;275(15):11465–9.

Lane RK, Hilsabeck T, Rea SL. The role of mitochondrial dysfunction in age-related diseases. Biochim Biophys Acta. 2015;1847(11):1387–400.

Lascaratos G, et al. Mitochondrial dysfunction in glaucoma: understanding genetic influences. Mitochondrion. 2012;12(2):202–12.

Leal-Esteban LC, Fajas L. Cell cycle regulators in cancer cell metabolism. Biochim Biophys Acta Mol Basis Dis. 2020;1866:165715.

Lee SH, et al. Expression of TGF-beta-induced matrix protein betaig-h3 is up-regulated in the diabetic rat kidney and human proximal tubular epithelial cells treated with high glucose. Kidney Int. 2003;64(3):1012–21.

Lee JH, et al. Effects of corneal neovascularization on the manifestations of Avellino corneal dystrophy (granular corneal dystrophy type II). Cornea. 2006;25(8):914–8.

Lee JJ, et al. Accessible chromatin structure permits factors Sp1 and Sp3 to regulate human TGFBI gene expression. Biochem Biophys Res Commun. 2011;409(2):222–8.

Levine B, Kroemer G. Autophagy in the pathogenesis of disease. Cell. 2008;132(1):27–42.

Lisch W, Seitz B. Lattice corneal dystrophy type 1: an epithelial or stromal entity? Cornea. 2014;33(10):1109–12.

Liu Y, et al. Gene Expression Profile of Extracellular Matrix and Adhesion Molecules in the Human Normal Corneal Stroma. Curr Eye Res. 2017;42(4):520–7.

Luo C, et al. Mitochondrial accumulation under oxidative stress is due to defects in autophagy. J Cell Biochem. 2013;114(1):212–9.

Maeng YS, Choi YJ, Kim EK. TGFBIp regulates differentiation of EPC (CD133(+) C-kit(+) Lin(−) cells) to EC through activation of the Notch signaling pathway. Stem Cells. 2015a;33(6):2052–62.

Maeng YS, et al. Histone methylation levels correlate with TGFBIp and extracellular matrix gene expression in normal and granular corneal dystrophy type 2 corneal fibroblasts. BMC Med Genomics. 2015b;8:74.

Maeng YS, et al. Inhibition of TGFBIp expression reduces lymphangiogenesis and tumor metastasis. Oncogene. 2016;35(2):196–205.

Maeng YS, et al. Role of TGFBIp in Wound Healing and Mucin Expression in Corneal Epithelial Cells. Yonsei Med J. 2017;58(2):423–31.

Martin C, Zhang Y. The diverse functions of histone lysine methylation. Nat Rev Mol Cell Biol. 2005;6(11):838–49.

Massague J. TGFbeta in cancer. Cell. 2008;134(2):215–30.

Mates JM, Perez-Gomez C, Nunez de Castro I. Antioxidant enzymes and human diseases. Clin Biochem. 1999;32(8):595–603.

Maynard CJ, et al. Accumulation of ubiquitin conjugates in a polyglutamine disease model occurs without global ubiquitin/proteasome system impairment. Proc Natl Acad Sci U S A. 2009;106(33):13986–91.

McLaren JW, et al. Keratocyte density: comparison of two confocal microscopes. Eye Contact Lens. 2005;31(1):28–33.

Mizushima N. Autophagy: process and function. Genes Dev. 2007;21(22):2861–73.

Mizushima N, et al. Autophagy fights disease through cellular self-digestion. Nature. 2008;451(7182):1069–75.

Munier FL, et al. Kerato-epithelin mutations in four 5q31-linked corneal dystrophies. Nat Genet. 1997;15(3):247–51.

Nam JO, et al. Regulation of tumor angiogenesis by fastatin, the fourth FAS1 domain of betaig-h3, via alphavbeta3 integrin. Cancer Res. 2005;65(10):4153–61.

Nam EJ, et al. Up-regulated transforming growth factor beta-inducible gene h3 in rheumatoid arthritis mediates adhesion and migration of synoviocytes through alpha v beta3 integrin: Regulation by cytokines. Arthritis Rheum. 2006;54(9):2734–44.

Norbury C, Nurse P. Animal cell cycles and their control. Annu Rev Biochem. 1992;61:441–70.

Ohno S, et al. RGD-CAP ((beta)ig-h3) enhances the spreading of chondrocytes and fibroblasts via integrin alpha(1)beta(1). Biochim Biophys Acta. 1999;1451(1):196–205.

Park SW, et al. Beta ig-h3 promotes renal proximal tubular epithelial cell adhesion, migration and proliferation through the interaction with alpha3beta1 integrin. Exp Mol Med. 2004;36(3):211–9.

Park SY, Jung MY, Kim IS. Stabilin-2 mediates homophilic cell-cell interactions via its FAS1 domains. FEBS Lett. 2009;583(8):1375–80.

Pattingre S, et al. Bcl-2 antiapoptotic proteins inhibit Beclin 1-dependent autophagy. Cell. 2005;122(6):927–39.

Poulsen ET, et al. Proteomic profiling of TGFBI-null mouse corneas reveals only minor changes in matrix composition supportive of TGFBI knockdown as therapy against TGFBI-linked corneal dystrophies. Febs j. 2018;285(1):101–14.

Rao NA, et al. Role of free radicals in uveitis. Surv Ophthalmol. 1987;32(3):209–13.

Rawe IM, et al. Beta-ig. Molecular cloning and in situ hybridization in corneal tissues. Invest Ophthalmol Vis Sci. 1997;38(5):893–900.

Runager K, et al. The insoluble TGFBIp fraction of the cornea is covalently linked via a disulfide bond to type XII collagen. Biochemistry. 2013;52(16):2821–7.

Schmitt-Bernard CF, et al. In vitro creation of amyloid fibrils from native and Arg124Cys mutated beta-IGH3((110–131)) peptides, and its relevance for lattice corneal amyloid dystrophy type I. Biochem Biophys Res Commun. 2000;273(2):649–53.

Schmitt-Bernard CF, et al. BIGH3 (TGFBI) Arg124 mutations influence the amyloid conversion of related peptides in vitro. Eur J Biochem. 2002;269(21):5149–56.

Shi Y, Massague J. Mechanisms of TGF-beta signaling from cell membrane to the nucleus. Cell. 2003;113(6):685–700.

Skonier J, et al. cDNA cloning and sequence analysis of beta ig-h3, a novel gene induced in a human adenocarcinoma cell line after treatment with transforming growth factor-beta. DNA Cell Biol. 1992;11(7):511–22.

Takacs L, et al. Immunohistochemical detection of betaIG-H3 in scarring human corneas. Graefes Arch Clin Exp Ophthalmol. 1999;237(7):529–34.

Vallabh NA, Romano V, Willoughby CE. Mitochondrial dysfunction and oxidative stress in corneal disease. Mitochondrion. 2017;36:103–13.

Wang DA, et al. Injury-elicited differential transcriptional regulation of phospholipid growth factor receptors in the cornea. Am J Physiol Cell Physiol. 2002;283(6):C1646–54.

Wenk J, et al. UV-induced oxidative stress and photoaging. Curr Probl Dermatol. 2001;29:83–94.

Yamada N, et al. Double mutation (R124H, N544S) of TGFBI in two sisters with combined expression of Avellino and lattice corneal dystrophies. Mol Vis. 2009;15:974–9.

Yellore VS, Rayner SA, Aldave AJ. TGFB1-induced extracellular expression of TGFBIp and inhibition of TGFBIp expression by RNA interference in a human corneal epithelial cell line. Invest Ophthalmol Vis Sci. 2011;52(2):757–63.

The Role of the Extracellular Matrix in *TGFBI*-Related Corneal Dystrophy Development

9

Ebbe Toftgaard Poulsen, Nadia Sukusu Nielsen, and Jan J. Enghild

Functions of the Human Cornea

The transparent nature of the cornea makes it unique compared to other tissues in the human body. Its primary function is to focus light on the retina with minimal scattering. In fact, the cornea is responsible for approximately 70% of the refractive power of the eye. The human cornea is about 0.55 mm thick and can be divided into five main layers consisting of the outermost epithelial cell layer, Bowman layer, stroma, Descemet membrane, and, lastly, the innermost cell layer called the endothelium. Each layer has distinctive properties. The epithelial cell layer is renewed continuously and, together with the tear film, is the first line of defense against corneal infections and foreign objects like dust and physical trauma. In addition, the epithelium, together with Bowman layer, is responsible for approximately 30% of corneal UV light absorption (Kolozsvari et al. 2002). The stroma occupies most of the corneal volume. It is a collagen-rich extracellular matrix (ECM) with a low density of cells (keratocytes). The stroma accounts for 70% of corneal UV light absorption.

Altogether, the human cornea absorbs nearly all UVB light (315–280 nm) and approximately one-third of UVA light (400–315 nm) entering the eye (Kolozsvari et al. 2002). UV light absorption by the three outermost layers is crucial for the protection of the non-regenerative endothelium on the posterior surface of the cornea, which ensures proper corneal hydration. Malfunction of the endothelial layer, as seen in Fuchs endothelial corneal dystrophy (FECD), leads to corneal edema and loss of vision (Liu et al. 2020).

The Architecture of the Corneal Stroma

The corneal stroma constitutes approximately 90% of the human cornea. The organization of the stromal collagen scaffold accounts for its transparency and mechanical strength. The regularly packed collagen fibrils consisting of collagen type I and V are organized into fibers, which form stacked layers, called lamellae. In the center of the cornea, collagen fibers run orthogonally to each other, whereas fibers in the corneal periphery branch interlace gradually into a circular orientation at the boundary between the cornea and sclera (Boote et al. 2008).

Other important components of the stromal matrix are the small leucine-rich proteoglycans (SLRP), decorin, biglycan, lumican, keratocan, and fibromodulin. The SLRPs help to maintain

E. T. Poulsen
The Danish Cornea Bank, Aarhus University Hospital, Aarhus, Denmark

N. S. Nielsen · J. J. Enghild (✉)
Department of Molecular Biology and Genetics, Aarhus University, Aarhus, Denmark
e-mail: jje@mbg.au.dk

E. K. Kim, R. D. Stulting (eds.), *TGFBI-related Corneal Dystrophies*,
https://doi.org/10.1007/978-981-96-0131-8_9

interfibrillar collagen spacing and are central to corneal hydration due to their glycosaminoglycan moieties. Several studies have demonstrated the vital functions of SLRPs in corneal development and collagen fibril assembly by examining murine corneas deficient in the SLRPs (Chen et al. 2015). Phenotypes of decorin—or biglycan-deficient mice have little impact on collagen fibrillogenesis, but mice deficient in both genes express a severe phenotype heavily affecting collagen fibril diameter and regularity throughout the stroma. Keratocan deficiency results in thinning of the corneal stroma and is, therefore, thought to be important for corneal hydration during development. Fibromodulin, which is only expressed during postnatal corneal development, is involved in cornea-sclera integration. Keratocan and fibromodulin deficiency does not affect the collagen fibril arrangement in the stroma. In contrast, lumican-deficient mice develop larger fibril diameters and an irregular fibril arrangement.

The last group of proteins important for the organization of corneal stromal collagen is the fibril-associated collagens with interrupted triple helices (FACIT), which includes collagen types IX, XII, XIV, XIX, and XXI. In the human cornea, all FACIT collagens have been identified (Dyrlund et al. 2012), but their function has not been fully characterized. The FACIT collagens do not form fibrils but are associated with the surface of the collagen fibrils in the ECM. Collagen type XI covalently binds the fibrils of type II collagen, whereas type XII FACIT collagen interacts with collagen type I (Ricard-Blum 2011). As FACIT collagens also interact with other proteins of the ECM, their binding to the collagen fibrils allows them to functionally modulate the surface properties of the collagen fibrils, including ECM-cell interactions.

The stromal ECM is maintained by keratocytes embedded within the collagen scaffold, which are responsible for the synthesis and secretion of ECM constituents. The keratocytes form a three-dimensional network via their filopodia allowing cell-cell communication (Chen et al. 2015). The compact cell body of keratocytes contains high levels of corneal crystallins that reduce light scattering by the cornea.

Overall, these unique properties and architecture of the cornea must be sustained throughout life for good vision to be maintained. Therefore, any imbalance in corneal homeostasis in any of the five layers can lead to visual impairment.

The Corneal Proteome

Despite the vast amounts of collagen in the cornea, many other proteins are important for maintaining homeostasis and structural integrity. The earliest proteomic study of the human cornea using mass spectrometry (MS) was completed in 2005, identifying 141 proteins of which 99 had not previously been associated with the cornea (Karring et al. 2005). The identified proteins were involved in many processes, including anti-angiogenesis, antimicrobial defense, protection from and transport of heme and iron, tissue protection against UV light radiation and oxidative stress, cell metabolism, and maintenance of intracellular and extracellular structures and stability.

As MS instrumentation improved, a total of 2070 distinct corneal proteins had been identified by 2011 (Galiacy et al. 2011). Pathway analyses of the identified proteins found redox and oxidative stress-related networks containing several aldehyde and alcohol dehydrogenases. Of these, aldehyde dehydrogenases 1 and aldehyde dehydrogenase 3A1, known as corneal crystallins, have UV light and oxidative protective roles in the corneal epithelium (Stagos et al. 2010). The pathway analysis also revealed a group of proteins putatively involved in corneal homeostasis, including structural proteins, proteins mediating cell-ECM contact, and proteolytic enzymes and their inhibitors.

The number of corneal proteins was further increased in a quantitative proteome study of the human cornea in 2012. In this study, 3250 proteins were identified, of which 1787 had not previously been associated with the cornea (Dyrlund et al. 2012), thus increasing the total number of corneal proteins identified by MS to 3708 in 2013 (Semba et al. 2013). For the first time, Dyrlund et al. examined the proteomes of the different layers of the cornea and quantitated the most abundant proteins within each layer. The relative amounts of various proteins in the different layers of the cornea reflected existing knowledge of protein constituents known for their roles in corneal anatomy and physiology.

The epithelium contained high amounts of cytoskeletal proteins such as keratins and dehydrogenases. The latter, as previously mentioned, are thought to function as corneal crystallins important for reducing the scattering of incoming light. The stroma was dominated by collagen type I, type IV, type V, type VI, and type XII and the SLRPs (except for fibromodulin). The endothelial layer, including Descemet membrane, showed large amounts of collagen type VIII and type IV, which are known to constitute the collagen scaffold of Descemet membrane (Tamura et al. 1991). Furthermore, the innermost layer also contained relatively high levels of the serine protease high-temperature requirement A1 (HtrA1), which is thought to be involved in protein quality control (Tennstaedt et al. 2012).

Strikingly, transforming growth factor beta-induced protein (TGFBIp) was among the 30 most abundant proteins in all three layers. While constituting only about 1% of the molar protein fraction in the epithelium and being the 21st most abundant protein, the molar fraction of TGFBIp in the other layers was substantially higher. In the stroma, it is the second most abundant protein after collagen type I, constituting approximately 17% of the molar protein fraction. In the endothelial layer including Descemet membrane, it is the most abundant protein constituting approximately 37% of the molar protein fraction (Dyrlund et al. 2012). Despite the high protein levels of TGFBIp throughout the cornea, its primary role remains a mystery.

Corneal TGFBIp Expression in Health and Disease

TGFBIp was first reported as a major constituent of the human cornea in 1994 (Escribano et al. 1994; Klintworth et al. 1994). Later, Nielsen et al. estimated the average concentration of TGFBIp to be 26.4 μM (±4.6) in the central part of human postmortem corneas from three individuals (age 81–91 years) (Nielsen et al. 2020). This high concentration of TGFBIp in human corneas appears to be an exception among animal species, as other species tend to display lower levels of TGFBIp. For instance, the protein level in mouse corneas has been estimated to be tenfold lower than that in human corneas (Poulsen et al. 2018). It is hypothesized that the higher level of TGFBIp in humans plays a role in the development of *TGFBI*-linked corneal dystrophies. However, two mouse models designed to mimic the phenotypes granular corneal dystrophy type 2(GCD2) and lattice corneal dystrophy (LCD) caused by the R124H and R124C mutations, respectively, have demonstrated that the murine TGFBIp level is sufficient for phenotype development (Kitamoto et al. 2020; Yamazoe et al. 2015).

Notably, a lack of phenotype development in the GCD2 mouse model was later shown to correlate with a halving of the murine TGFBIp protein level, which strongly supports a correlation between the concentration of TGFBIp in the cornea and phenotype development of *TGFBI*-related dystrophies (Lukassen et al. 2020). The level of TGFBIp changes over time in the normal cornea. Karring et al. found TGFBIp levels in the human cornea to increase by 30% from the first to the second decade of life (Karring et al. 2010). It is not known whether this is due to increased synthesis in the cornea or an accumulation of TGFBIp over time.

In addition to the cornea, TGFBIp is widely expressed in various tissues of the human body including the heart, breast, testes, prostate, intestines, skin, ovary, liver, pancreas, lung, leukocytes, placenta, and kidney (Nielsen et al. 2020). Despite its prevalence throughout the human body, disease-causing mutations in TGFBIp only seem to affect the corneal tissue as demonstrated by a systemic investigation of a patient suffering from LCD (El Kochairi et al. 2006). The study showed that TGFBIp accumulation was confined to the cornea and no other tissue or organ in the patient. This suggests that the corneal environment is unique when compared to the other

TGFBIp-containing tissues. Observation of clear zones surrounding blood vessels in the corneas of patients with GCD2 has led to the hypothesis that avascularity makes the cornea uniquely susceptible to the clinical effects of mutations in TGFBIp (Han et al. 2016).

TGFBIp is believed to be synthesized by the corneal epithelium and then secreted into the stroma (Escribano et al. 1994). This hypothesis is supported by in situ hybridization analysis of mouse corneas, which reveals major transcript levels solely in the epithelium. Keratocytes also express TGFBIp in the stromal periphery (Liu et al. 2017) and during wound healing (Takacs et al. 1999). Interestingly, despite TGFBIp being abundant in the endothelium and Descemet membrane (Dyrlund et al. 2012), it does not appear to be synthesized by the endothelial cell layer. The fact that deposits reappear in the cornea of *TGFBI*-linked corneal dystrophy patients after penetrating keratoplasty strongly supports the concept that the majority of corneal TGFBIp originates from the epithelium, as the recipient epithelium will repopulate the donor cornea shortly after transplantation (Han et al. 2016).

The Primary Structure of TGFBIp

TGFBIp is composed of 683 amino acid residues, including a 23 amino acid N-terminal secretory signal peptide. The protein consists of five domains including an N-terminal cysteine-rich domain of periostin and TGFBIp (CROPT) domain followed by four fasciclin 1 (FAS1) domains (FAS1-1, FAS1-2, FAS1-3, and FAS1-4). The C-terminal region (residues 633–683) following the last FAS1 domain contains an RGD integrin-binding motif. The major part of TGFBIp in the human cornea appears C-terminally truncated. The truncation has been shown to occur mainly at A657, just three residues downstream of the RGD integrin-binding motif. Whether this proteolytic event occurs before or after secretion is currently unknown, but its proximity to the RGD motif has been suggested to regulate the cell-binding properties of TGFBIp (Andersen et al. 2004). TGFBIp contains 11 cysteine residues, of which six are located in the CROPT domain (Lukassen et al. 2016). Ten of the 11 cysteines participate in the formation of inter—and intra-domain disulfide bonds tethering together the domains in TGFBIp. The last cysteine residue not participating in an intramolecular disulfide bond is located in the CROPT domain at residue position C65 and is believed to be involved in anchoring TGFBIp to the corneal stroma (Lukassen et al. 2016).

Interaction of TGFBIp with the Corneal ECM

An extensive characterization of TGFBIp as an integrin-binding protein has been carried out over the years. In addition to the RGD motif situated in the C-terminal, several other integrin-binding sites have been found throughout the protein. The role of TGFBIp as an integrin-binding protein has been described in detail in the previous chapter and will not be the focus of this chapter. However, TGFBIp has also been reported to bind several ECM proteins, of which many are well known for their role in the proper arrangement of the collagen scaffold in the stroma and the Descemet's membrane (Table 9.1).

As described at the beginning of this chapter, SLRPs like biglycan and decorin are major con-

Table 9.1 ECM proteins interacting with TGFBIp

Protein	References
Bigycan	Reinboth et al. (2006)
Decorin	Reinboth et al. (2006)
Fibronectin	Billings et al. (2002), Kim et al. (2002)
Laminin	Kim et al. (2002)
Periostin	Kim et al. (2009)
Collagen type I	Billings et al. (2002), Hashimoto et al. (1997), Kim et al. (2002)
Collagen type II	Hashimoto et al. (1997), Kim et al. (2002)
Collagen type IV	Hashimoto et al. (1997)
Collagen type VI	Hanssen et al. (2003), Kim et al. (2002), Rawe et al. (1997), Reinboth et al. (2006)
Collagen type XII	Runager et al. (2013)

tributors to stromal integrity. In vitro experiments have demonstrated the binding of TGFBIp to the core protein of both biglycan and decorin, with a preference for the former (Reinboth et al. 2006). These studies also showed that TGFBIp could form high molecular weight complexes with collagen type VI with and without the presence of any of the two SLRPs. Hence, TGFBIp is thought to play a role in modulating the aggregation of collagen type VI with SLRP.

TGFBIp has also been shown to interact with fibronectin, an abundant ECM component involved in cell attachment and spreading, wound healing, and vascularization (Kim et al. 2002). The specific binding was shown to involve the collagen/gelatin binding domain of fibronectin (Billings et al. 2002). In addition, TGFBIp interacts with laminin (Kim et al. 2002) a basement membrane protein that is abundant in Descemet membrane and is implicated in cell adhesion, migration, and differentiation.

Five of the collagens have been identified as TGFBIp interaction partners, including both fibrillar and FACIT collagens. Collagen types I, II, and IV are reported as noncovalent interaction partners of TGFBIp (Billings et al. 2002; Hashimoto et al. 1997; Kim et al. 2002), whereas noncovalent and covalent interactions have been found with collagen type VI (Hanssen et al. 2003; Kim et al. 2002; Rawe et al. 1997; Reinboth et al. 2006). In the study by Rawe et al., TGFBIp and collagen type VI were copurified from rabbit corneas but required a reduction to enter the SDS-PAGE gel (Rawe et al. 1997). This was corroborated by findings of Hanssen et al., showing that collagen type VI microfibrils purified from nuchal ligament tissue were covalently bound to TGFBIp through a reducible bond (Hanssen et al. 2003). In the human cornea, more than half of TGFBIp is bound covalently to the insoluble ECM fraction through a reducible bond (Andersen et al. 2004). Later, collagen type XII was identified as a major covalent interaction partner of TGFBIp in the cornea. The binding site was suggested to involve the NC3 domain of collagen type XII (Runager et al. 2013). The residue in TGFBIp responsible for forming the reducible bond to the ECM is believed to be cysteine 65 situated in the CROPT domain. The C65 residue does not participate in the intra-disulfide assembly of TGFBIp (Lukassen et al. 2016) and would be available for covalent interaction with collagen types VI and XII.

The interaction between TGFBIp and ECM proteins crucial for the proper assembly of the collagen scaffold in the stroma could be indicative of a similar structural importance of TGFBIp for corneal integrity. However, the structure of corneas of TGFBIp-deficient mice was not compromised significantly, despite the fact that collagen type VI and type XII were regulated differently than those of WT mice (Poulsen et al. 2018). This suggests that TGFBIp might have a role beyond functioning as a structural ECM protein in the cornea.

Processing of TGFBIp in the Normal Cornea

It is generally believed that the major fraction of TGFBIp in the cornea originates from the epithelium, but little is known about the clearance mechanism of TGFBIp. Studies have shown that extracellular degradation to some extent is involved in the turnover of corneal TGFBIp. When analyzed by two-dimensional gel electrophoresis (2DE) followed by immunoblotting, several TGFBIp isoforms with a molecular weight in the 30–60 kDa range and an isoelectric point between 5.5 and 6.2 appear and form a characteristic "zigzag" pattern (Fig. 9.1) (Karring et al. 2010). The heterogeneous nature of corneal TGFBIp is a product of posttranslational modifications, more specifically post-processing by yet unknown proteases suggested to involve aminopeptidases (Karring et al. 2010). MS analysis has shown that the characteristic "zigzag" pattern occurs because the N-terminal of the mature protein has been trimmed. In addition to the "zigzag" pattern, the smallest isoform observed on a 2D-PAGE of corneal extract spans the FAS1-4 domain of TGFBIp migrating with a molecular weight of 14 kDa (Karring et al. 2010). The degree of TGFBIp processing appears to be different in postnatal corneas than it is for the

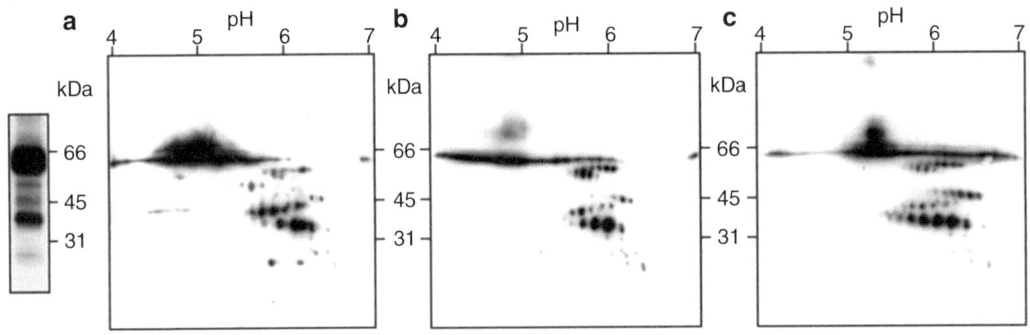

Fig. 9.1 Differential processing of TGFBIp in the cornea through postnatal development and aging. 2DE immunoblots of corneal TGFBIp from (**a**) a 6-month-old (**b**), a 1-year-old, and (**c**) a 6-year-old individual. The last immunoblot is representative of individuals at the age of 6–86 years. A characteristic "zigzag" pattern is observed on all the immunoblots, but the relative abundance of the different TGFBIp isoforms changes with age. To the left of panel (**a**), a 1D immunoblot shows the detection of the lower molecular weight isoforms of corneal TGFBIp from a 14-year-old individual when the blot is developed for a longer period of time. Adapted from Karring et al. (2010)

remaining lifetime of the human, suggesting that TGFBIp may be involved in the maturation of the infant cornea (Karring et al. 2010). As previously mentioned, the corneal TGFBIp concentration increases by 30 percent during the first two decades of life. The increase in the TGFBIp level may relate to the time of onset of phenotype in *TGFBI*-linked corneal dystrophies, as protein deposits rarely appear in early infancy, further supporting a concentration dependency for phenotype development.

Few other studies have addressed posttranslational modification of TGFBIp. Only phosphorylation at S37 has been reported (Bian et al. 2014; Srivastava and Srivastava 1999), and no evidence for other posttranslational modification was observed in a study on TGFBIp from corneal extracts (Andersen et al. 2004).

As previously mentioned, it is remarkable that the deposition of TGFBIp appears to be specific to the cornea, even though TGFBIp is widely expressed in many tissues (Nielsen et al. 2020). A unique extracellular proteolytic environment in the cornea may, therefore, nurture the disease progression of *TGFBI*-linked corneal dystrophies. Comparing 2D immunoblots of the in vivo TGFBIp degradation pattern in other human specimens including the skin, plasma, and platelets with those of the cornea clearly demonstrates differential processing. Though extensively processed, skin TGFBIp is confined to a smaller isoelectric point range and differs in its degradation pattern. Plasma and platelet TGFBIp show only a few degradation products compared to those of the cornea (Nielsen et al. 2020). In addition to being tissue-specific, processing of corneal TGFBIp may also be species-specific. 2DE immunoblots of wild-type mouse corneas and corneas of the humanized R124H mouse model show highly similar in vivo degradation patterns of mouse and human TGFBIp. Only two faint low molecular spots were unique to the humanized mouse cornea (Lukassen et al. 2020). However, mouse corneas showed a less pronounced TGFBIp degradation "zigzag" pattern when compared to the human cornea, and trimmed TGFBIp isoforms were shifted toward a lower isoelectric point than observed in human corneas (Lukassen et al. 2020). This may be due to different TGFBIp processing by different extracellular proteases in the mouse and human. Indeed, the generation of a mouse model expressing the endogenous mouse TGFBIp containing the R124C mutation normally resulting in amyloid deposition in humans with LCD displayed the GCD phenotype instead (Kitamoto et al. 2020). As discussed by the authors, this shift in phenotype may be due to a difference in the proteolytic machinery of the mouse cornea. The differential processing observed between mice and humans may be due to the anatomical differences between corneas of the two species but may also

Table 9.2 TGFBI-linked cases analyzed by laser capture microdissection and mass spectrometry

Genotype	Phenotype	Characteristics	References
R124C	LCD1	Stromal and Bowman's amyloid deposits	Courtney et al. (2015)
R124H	GCD2	Stromal amorphous aggregates	Karring et al. (2012)
A546D	LCD variant	Stromal amyloid deposits	Karring et al. (2013)
A546D/P551Q	LCD variant	Stromal amyloid deposits	Poulsen et al. (2014b)
R555W	GCD1	Stromal amorphous aggregates	Courtney et al. (2015)
V624M	LCD variant	Stromal amyloid deposits	Karring et al. (2012)
H626R	LCD variant	Stromal amyloid deposits	Venkatraman et al. (2017)

rely on alteration in the proteolytic machinery. Proteases represent the largest group of enzymes in humans, consisting of at least 553 proteases. Correspondingly, the number of proteases in mice is at least 628 (Puente et al. 2003) leaving the possibility for species-specific processing. Of the total proteins identified in human cornea so far (Semba et al. 2013), 97 are classified as active proteases accordingly to the Gene Ontology and the MEROPS databases (Ashburner et al. 2000; Rawlings et al. 2018). Of these, 58 are categorized to be within the extracellular space and thereof potentially involved in the N-terminal processing of TGFBIp in the human cornea. Alternatively, they may become involved in mutant TGFBIp proteolysis as disease-causing mutations affect the structure or function of the protein.

Evidence for the importance of the proteolytic turnover of TGFBIp in the development of *TGFBI*-linked corneal dystrophies was obtained by comparison of 2DE of the normal human cornea to corneal tissue of patients genotyped with the R124C, R124H, and R124L transforming growth factor beta-induced (*TGFBI*) mutations (Korvatska et al. 2000). These mutations represent the LCD, GCD2, and RBCD phenotypes, respectively. While the R124C mutation resulted in a major accumulation of a 44 kDa N-terminal TGFBIp fragment, R124H and R124L displayed an accumulation of intact or near-intact TGFBIp. The 2DE further showed differential degradation of patient TGFBIp compared to normal cornea affecting the location and number of degradation spots identified on the 2DE. TGFBIp secreted by cultivated fibroblasts established from R124C patients does not match the degradation pattern observed for R124C corneal tissue. This strongly implies that the aberrant TGFBIp processing observed between the R124 genotypes is confined to the ECM of the cornea (Korvatska et al. 2000).

Recently, several studies have confirmed the involvement of altered TGFBIp turnover by using MS analyses on patient tissue. Of the 74 disease-causing mutations reported so far in humans, 7 genotypes have been investigated, representing both GCD and LCD phenotypes (Table 9.2) (Courtney et al. 2015; Karring et al. 2012, 2013; Poulsen et al. 2014b; Venkatraman et al. 2017). In these studies, laser capture microdissection was elegantly used to isolate corneal protein deposits followed by protein profiling using MS. These analyses have clearly distinguished among GCD and LCD at the molecular level.

Protein Profiling of Deposits in *TGFBI*-Linked Corneal Dystrophy

The *TGFBI*-related corneal dystrophies display a dual aggregation propensity that resides within TGFBIp, resulting in either amorphous (GCD) or amyloid (LCD) insoluble protein aggregates upon the introduction of a disease-causing mutation. The formation of insoluble protein structures is a well-known hallmark of many diseases, which generally follows two different aggregation pathways leading to either amorphous aggregates or highly ordered amyloid structures (Ramshini et al. 2011). While larger proteins destined for aggregation tend to follow the amorphous pathway involving the intact protein, the amyloidogenic pathway often involves smaller proteins or peptides produced by abnormal proteolytic processing of intact proteins. As will be discussed below in detail, this generalization also seems to apply to TGFBIp. While amorphous aggregates associated with the GCD phenotype consist of largely intact TGFBIp deposits, TGFBIp appears to be extensively processed in the amyloid aggregates associated with the LCD phenotype.

Characterization of GCD Deposits

The first article characterizing protein deposits of *TGFBI*-linked corneal dystrophy tissue using laser capture microdissection (LCM) followed by MS analysis was published in 2012 by Karring and colleagues (Fig. 9.2) (Karring et al. 2012). Using formalin-fixed paraffin-embedded corneal tissue, the authors compared the protein composition of the aggregates with that of healthy control tissue to identify the accumulating proteins. This defined the technical approach applied during all of the following studies performing protein composition analyses on *TGFBI*-linked corneal dystrophy tissue (Courtney et al. 2015; Karring et al. 2013; Poulsen et al. 2014b; Venkatraman et al. 2017).

Two studies characterizing the protein profiles of GCD1, caused by the R555W mutation, and GCD2, caused by the R124H mutation, have provided insight into the GCD phenotype. In the latter case, intact or near-intact TGFBIp was shown to constitute most of the amorphous aggregates in GCD2 tissue. This finding corroborated the work of Klintworth and coworkers, who analyzed corneal GCD2 tissue not subjected to microdissection (Klintworth et al. 1998). The same tendency of intact TGFBIp aggregation was later observed in GCD1 tissue (Courtney et al. 2015). These studies collectively showed that, despite a considerable distance in the primary sequence, GCD caused by a mutation either in the FAS1-1 or FAS1-4 domain of TGFBIp resulted in corneal aggregates of intact protein and with a similar phenotypic outcome. In addition to the distinct accumulation of TGFBIp in GCD deposits, the protein profile reported by the two studies resembled healthy corneal stroma to a high degree. The study of GCD1 tissue only identified three unique proteins associated with GCD1 deposits not present in control tissue, including hemoglobin subunit beta, actin cytoplasmic 1, and desmoplakin (Courtney et al. 2015). In the GCD2 study, serum amyloid P (SAP), type III collagen, clusterin, and histone H3-like protein were enriched in GCD2 aggregates (Karring et al. 2012).

The finding of SAP in GCD2 aggregates is noteworthy, as SAP usually associates with amyloid deposits of various misfolding diseases (Pepys et al. 1979) in addition to being a multiple protein-binding protein (Poulsen et al. 2017). Contrary to this, SAP was not identified either in the GCD1 tissue analyzed (Courtney et al. 2015) or in GCD2 deposits unintentionally generated as a consequence of laser-assisted in situ keratomileusis (LASIK) surgery on heterozygous patients

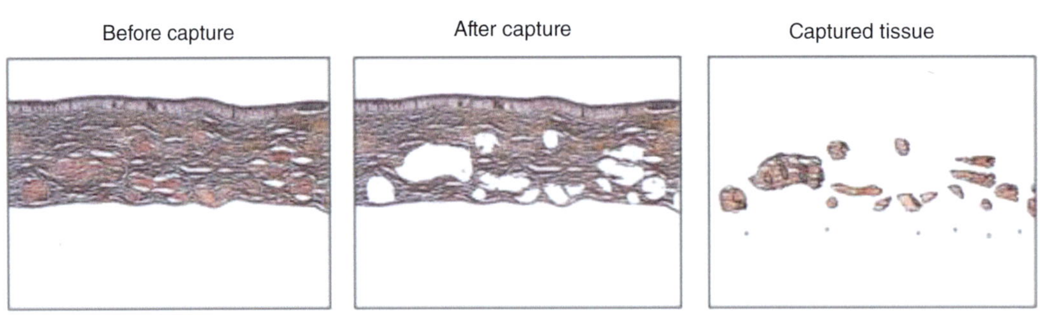

Fig. 9.2 Amyloid deposits removed by LCM from fixed corneal tissue of an LCD patient with the A546D/P551Q double mutation in *TGFBI*. The tissue was stained with Congo red to highlight amyloid deposition. From the left is depicted the tissue before LCM, in the middle the tissue after LCM, and to the right corneal amyloid deposits isolated and used for MS analysis. The photographs were taken with a light microscope at 4× magnification. Adapted from Poulsen et al. (2014a, 2014b)

(Poulsen et al. 2016). As a generic biomarker for amyloid depositions, the presence of SAP in GCD2 aggregates may indicate an underlying amyloid deposition masked by the amorphous aggregates. This would correlate with the characteristic mixed phenotype observed in some instances for GCD2, in which opacities that appear to be amyloid are observed in later stages of the disease.

Characterization of LCD Deposits

While the GCD analyses showed some resemblance to healthy stromal tissue, a distinct protein profile defines the LCD deposits. Thus, the major common denominator between LCD and GCD deposits appears to be the protein deposition of mutant TGFBIp.

Patients carrying the *TGFBI* mutations R124C, A546D, A546D/P551Q, V624M, or H626R have been characterized (Courtney et al. 2015; Karring et al. 2012, 2013; Poulsen et al. 2014b; Venkatraman et al. 2017). These studies showed a strong resemblance in deposited proteins despite the mutation being situated in the FAS1-1 or FAS1-4 domain of TGFBIp. This is similar to the characteristics of other amyloid-binding profiles, which suggests that the location of amyloid deposits might have a greater influence on the composition of the amyloid-associated proteins than does the exact structure of the amyloidogenic fibrillar aggregates (Juhl et al. 2019).

In addition to the accumulation of TGFBIp, several other amyloidogenic proteins, as well as nonfibrillar amyloid-associated proteins, are found associated with the amyloid deposits. The amyloidogenic proteins include apolipoprotein A-I, apolipoprotein A-IV, lysozyme C, protein S100-A8, protein S100-A9, and lactotransferrin. It is unknown whether the amyloidogenic proteins are directly implicated in the formation of deposits along with TGFBIp or merely figure in the deposits as innocent coaggregating bystanders. The nonfibrillar amyloid-associated proteins involve SAP, clusterin, apolipoprotein E, apolipoprotein D, prostaglandin-H2 D-isomerase, complement component C9, decorin, and cystatin-A.

Of note, several proteins involved in proteolysis have been identified in the protein profiles of amyloid deposits (Courtney et al. 2015; Karring et al. 2012, 2013; Poulsen et al. 2014b; Venkatraman et al. 2017). Most profound was the identification of HtrA1, which associates with all analyzed LCD cases. In addition, kallikrein-14 was identified in the A546D deposits (Karring et al. 2012), whereas carboxypeptidase B2, caspase-14, cathepsin D, cathepsin L2, and plasminogen were identified solely in deposits of A546D/P551Q patients (Poulsen et al. 2014b). Furthermore, the protease inhibitor cystatin-A was identified in amyloid deposits of A546D, A546D/P551Q, and R124C cases, while glia-derived nexin, a serpin inhibitor, was only identified in A546D deposits. The metalloprotease inhibitors 1 and 3 and serpin B12 were identified in A546D/P551Q amyloid deposits. Collectively, the presence of several proteases and inhibitors within the amyloid deposits supports extensive proteolytic activity in LCD progression possibly involved in the aberrant TGFBIp turnover originally reported by Korvatska and colleagues (Korvatska et al. 2000). Contrary to this, GCD deposits did not display any proteins involved in proteolysis, which supports the existence of two very different disease mechanisms (Fig. 9.4).

A thorough examination of the TGFBIp-derived peptides observed in amyloid deposits has shown extended accumulation of various ragged isoforms of two polypeptide regions in FAS1-4 restricted to a disease-causing mutation in the FAS1-4 domain of TGFBIp (Karring et al. 2012, 2013; Poulsen et al. 2014b; Venkatraman et al. 2017). The two polypeptide regions have been designated as the F515-R533 and Y571-R588 peptides in the literature, referring to the two peptides obtained after the enzymatic cleavage of TGFBIp by trypsin, which normally is used in sample preparation for mass spectrometry analyses. Both peptides can form in vitro amyloid fibrils capable of seeding amyloid fibrillation of the FAS1-4 domain carrying the LCD-linked A546T mutation. Furthermore, the synthetic amyloid fibrils were shown to coprecipitate in a non-amyloid manner endogenous TGFBIp isolated from human cornea

(Sørensen et al. 2015). This suggests that additional aggregation events in the cornea may be initiated by the formation of amyloid deposits.

The TGFBIp peptide distribution observed in amyloid deposits due to the FAS1-1 R124C mutation showed a less pronounced accumulation of the F515-R533 and Y571-R588 peptide regions (Courtney et al. 2015). Instead, the L128-R172 TGFBIp peptide was found elevated in amyloid aggregates when compared to levels in healthy stroma. This supports the notion that TGFBIp participation in amyloid formation relies on more than one pathway depending on the site of mutation, all showing the same phenotype in the cornea.

Examining the endogenous cleavage sites observed for degraded TGFBIp in amyloid deposits showed enzymatic preferences for cleavage after the aromatic and hydrophobic amino acid residues (A, F, H, I, L, M, V, W, and Y), as well as the hydrophilic amino acids T, N, Q, D, and E (Courtney et al. 2015; Karring et al. 2012, 2013; Poulsen et al. 2014b; Venkatraman et al. 2017). The cleavage sites match the specificity of the serine protease HtrA1 (Truebestein et al. 2011), found associated with all LCD cases analyzed so far. This supports the concept that HtrA1 is involved in the in vivo processing of TGFBIp in the cornea.

The Involvement of HtrA1 in LCD Progression

The human serine protease HtrA1 is best known for its involvement in various types of cancer (Zurawa-Janicka et al. 2010), age-related macular degeneration (Vierkotten et al. 2011), and Alzheimer's disease (Grau et al. 2005), but it has also been associated with less common diseases such as immunoglobulin light chain amyloidosis of the eye (Nielsen et al. 2014). The identification of HtrA1 in all the LCD cases analyzed so far suggests a role for HtrA1 in disease progression or in removing protein aggregates from the cornea. Typically, protein misfolding and deposition are involved in many of the diseases associated with HtrA1. Therefore, it is generally accepted that HtrA1 plays a role in protein quality control, similar to that of the bacterial homologs (Runyon et al. 2007).

HtrA1 is a homotrimeric nonglycosylated serine protease belonging to the HtrA family encompassing four proteases in humans. The 51 kDa monomeric subunit is composed of four domains including the N-terminal IGFBP-like and Kazal-like domains, a central catalytic domain, and a C-terminal PDZ domain. The PDZ domain is believed to mediate substrate recognition (Runyon et al. 2007) and to facilitate binding to amyloid fibrils (Poepsel et al. 2015). The roles of the two N-terminal domains are still unknown; however, they can be removed from the remaining HtrA1 protease by redox-mediated autolysis in the presence of the redox-protein thioredoxin (Risør et al. 2014). Thioredoxin has been associated with two LCD cases and may therefore influence HtrA1 functionality in the cornea (Karring et al. 2013; Poulsen et al. 2014b). However, HtrA1 specificity or its activity does not appear to be affected by autolysis (Poulsen et al. 2019), and the functional significance remains elusive.

In addition to the endothelium and Descemet membrane (Dyrlund et al. 2012; Poulsen et al. 2014a), intact HtrA1 has also been found in the healthy corneal stroma (Poulsen et al. 2019). Furthermore, HtrA1 was enriched in peri-amyloid tissue (stromal tissue from LCD patients without apparent protein deposits) compared to control tissue in the studies characterizing deposits from A546D and H626R corneas (Karring et al. 2013; Venkatraman et al. 2017). HtrA1 has been ascribed a role in disintegrating the amyloid structure in vitro (Poepsel et al. 2015). Thus, HtrA1 may attempt to remove misfolded TGFBIp protein in the prefibrillar stage or disintegrate preformed amyloid fibril structures followed by degradation of the disintegrated protein.

Poulsen and coworkers confirmed the role of mutant TGFBIp and its FAS1-4 domain as HtrA1 substrates by studying the in vitro proteolysis of wild-type and mutant TGFBIp representing both the GCD and LCD genotypes (Poulsen et al. 2019). In agreement with the in vivo studies on protein deposits, HtrA1 showed the highest preferences for mutant TGFBIp associated with the LCD phenotype, especially when using the FAS1-4 domain as a substrate. Degraded FAS1-4 were capable of forming amyloid fibrils consisting mainly of peptides corresponding to the Y571-R588 region highly represented in in vivo LCD deposits. Furthermore, 2DE immunoblots of LCD corneal tissue having the A546D or V624M TGFBIp mutation clearly showed a reduction in TGFBIp degradation products, not seen for GCD2 tissue (Poulsen et al. 2019). Collectively, this suggests that HtrA1 in an attempt to proteolyze misfolded TGFBIp generates TGFBIp peptides prone to amyloid fibrillation and, therefore, maybe a driving force in the progression of LCD. HtrA1 is only associated with the LCD phenotype, suggesting that LCD-linked mutations are the key to unlock the HtrA1-mediated degradation of TGFBIp.

Structure of Wild-Type TGFBIp

The crystallization of TGFBIp was accomplished in 2009 (Runager et al. 2009) and characterized by X-ray crystallography in 2017 (Fig. 9.3) (Garcia-Castellanos et al. 2017). The structure might contribute to the understanding of the functions of wild-type TGFBIp and the effects of mutations associated with *TGFBI*-linked corneal dystrophies (Garcia-Castellanos et al. 2017). The

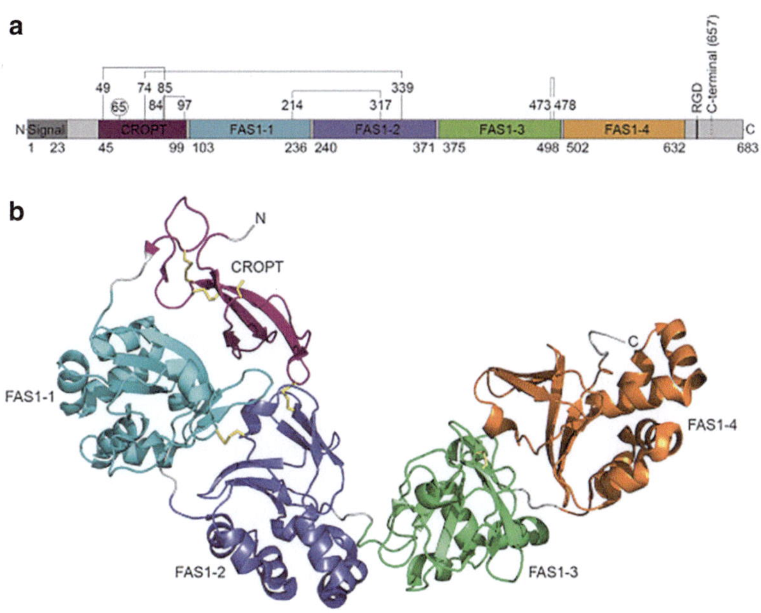

Fig. 9.3 TGFBIp schematic representation and three-dimensional structure. Panel (**a**) shows a schematic representation of the secreted form of TGFBIp (24–683) with the positions of the disulfide bonds, the RGD motif, and an N-terminal truncation site at residue 657. TGFBIp contains two intra-domain disulfide bridges in the CROPT domain (C49–C85 and C84–C97) and one in the FAS1–3 domain (C473–478). One inter-domain disulfide bridge is found between the CROPT domain and the FAS1–2 domain (C74–C339) and another between the FAS1–1 domain and the FAS1–2 domain (C214–C317). C65 is not engaged in any disulfide bridge. Panel (**b**) depicts the three-dimensional structure of TGFBIp (residues 43–637) as a ribbon plot with the CROPT domain (magenta) and the four FAS1 domains, FAS1–1 (cyan), FAS1–2 (blue), FAS1–3 (green), and FAS1–4 (orange). The free C65 and the disulfide bridges are shown as yellow sticks. The structure of flexible regions like the N—and C-terminus of TGFBIp could not be determined. Adapted from Nielsen et al. (2020)

overall shape of TGFBIp resembles that of an elongated banana that approximately spans 120 Å, with the FAS1 domains arranged like pearls on a string. Flexible regions like the N-terminal (residues 24–42) and C-terminal (residues 638–657) containing the RGD motif are not observed in the structure. The CROPT domain consists of a central three-stranded β-sheet and is cross-connected by two intra-domain disulfide bonds (C49–C85 and C84–C97) and one inter-domain disulfide bond (C74–C339) that connects it to the FAS1-2 domain. In addition, the free cysteine (C65) is located on the surface at the beginning of a flexible loop making it accessible for binding to type XII collagen (Runager et al. 2013). The four FAS1 domains consist of a mixture of α-helices and β-sheets and conform to the same general architecture (Garcia-Castellanos et al. 2017). The FAS1-3 domain contains an intra-domain disulfide bond (C473–C478), and the FAS1-1 and FAS1-2 domains are connected by an inter-domain disulfide bond (C214–C317). The structure can be divided into a left moiety composed of the CROPT, FAS1-1, and FAS-1-2 domains and a right moiety containing the FAS1-3 and FAS1-4 domains. The left moiety is more compact compared to the right moiety due to the two inter-domain disulfide bonds connecting the FAS1-2 domain with the CROPT and FAS1-1 domain, respectively. The interface between the two moieties is flexible and may act as a hinge, enabling different conformations of the whole protein that could be linked to the function of TGFBIp.

TGFBIp does not have a well-defined quaternary structure. However, dimers, trimers, and higher-order multimers have been observed in several studies (Basaiawmoit et al. 2011; Kim et al. 2002). In a small-angle X-ray study of recombinant human TGFBIp, a concentration-dependent trimerization of TGFBIp was observed (Basaiawmoit et al. 2011). Considering the high concentration of TGFBIp in the cornea, it cannot be ruled out that dimers and/or trimers of TGFBIp are physiologically relevant and that these oligomeric structures might play a role in the propensity of TGFBIp to aggregate.

Biophysical and Structural Effects of Disease-Causing Mutations in the FAS1-4 Domain

Mutations associated with *TGFBI*-linked corneal dystrophies are mainly located in the FAS1-1 and FAS1-4 domain. They are, however, most frequently observed in the FAS1-4 domain, which suggests that the FAS1-4 domain might be important for the stability of the full-length protein. This hypothesis is supported by information obtained from the structure of wild-type TGFBIp. In contrast to the intermolecular interactions of the CROPT, FAS1-1, and FAS1-2 domains, only a small interface between the FAS1-3 and FAS1-4 domain is evident, making FAS1-4 more sensitive to mutation-induced destabilizing effects, as these cannot be offset by stabilizing inter-domain contacts. Consistent with these observations, the FAS1-4 domain is more sensitive to trypsin cleavage than the other three FAS1 domains, indicating that it is the most conformationally unstable domain of TGFBIp (Runager et al. 2011).

Several studies have shown that disease-causing mutations in the FAS1-4 domain affect the stability of the isolated FAS1-4 domain and the full-length protein similarly (Andreasen et al. 2012; Koldsø et al. 2015; Nielsen et al. 2017; Poulsen et al. 2019; Runager et al. 2011). In general, mutations associated with LCD destabilize the isolated FAS1-4 domain and full-length TGFBIp, whereas mutations associated with GCD1 and TBCD stabilize or do not change the stability of the FAS1-4 domain and full-length protein. This link between the clinical phenotype and the impact on protein folding and stability has been extensively investigated in a study encompassing 30 FAS1-4 mutants, spanning mutations associated with both LCD, GCD1, variant GCD2, and TBCD (Stenvang et al. 2018). Recombinantly expressed FAS1-4 domain mutants were studied with a wide range of biophysical methods. All LCD mutations were destabilizing, and 14 of the 22 LCD mutants could not be purified, illustrating their significant effects on FAS1-4 stability. In addition, two variant GCD2 mutants were also shown to be destabilizing. Interestingly, all of the LCD mutations and the two GCD2 mutations occur in buried positions when grouped on the FAS1-4 structure, which explains their effect on stability (Garcia-Castellanos et al. 2017; Stenvang et al. 2018). However, the LCD mutations are not predicted to increase the aggregation propensity of FAS1-4 (Stenvang et al. 2018). Hence, the LCD mutations induce aggregation because they increase the population of unfolded FAS1-4, not because the unfolded state becomes more aggregation-prone due to the mutation. Consistently, only very few LCD-associated mutations are situated in the sequence Y571-R588 that has been identified as the core of TGFBIp fibril deposits (Sørensen et al. 2015). Most likely, the consequence of the LCD-linked mutations is the loss of structural integrity of the FAS1-4 domain. The compromised protein may hereof become a substrate for proteases such as HtrA1 as previously discussed, allowing degradation products to aggregate.

It seems counterintuitive that aggregation of TGFBIp is promoted by the near-wild-type stability of R555Q (TBCD) and a stabilizing mutation such as R555W (GCD1) (Runager et al. 2011; Stenvang et al. 2018; Underhaug et al. 2013). However, a bioinformatics analysis suggests that these mutations reduce the solubility of the FAS1-4 domain (Stenvang et al. 2018). The charged residue R555 is found at a relatively exposed site, and replacement with a large aromatic side chain such as tryptophan or a polar residue like glutamine is predicted to increase the protein's tendency to precipitate (Stenvang et al. 2018). In contrast, the variant GCD2 mutations decrease stability rather than solubility. The two mutations M619K and L550P may not be compatible with amyloid formation and lead to a faster non-amyloid aggregation pathway. Hence, the GCD1 and TBCD mutations can influence the nature of the aggregate to a greater extent than LCD mutants can, as these simply increase the tendency to unfold.

Biophysical and Structural Effects of Disease-Causing Mutations in the FAS1-1 Domain

Only 14 of the 74 reported mutations associated with *TGFBI*-linked corneal dystrophies involve the FAS1-1 domain (Nielsen et al. 2020). A few studies have investigated the impact of some of these FAS1-1 mutations on the protein folding and stability of TGFBIp.

Unlike the LCD-linked mutations in the FAS1-4 domain, the LCD1-associated mutation R124C in the FAS1-1 domain does not affect the stability of the full-length protein (Runager et al. 2011). This correlates well with the fact that R124 is surface exposed in the structure of wild-type TGFBIp and would not be expected to affect the protein fold. In addition, the accumulation of the FAS1-4 amyloidogenic core region (Y571–588) is less prominent in plaque tissue from LCD1 patients (Courtney et al. 2015). Instead, a 22 amino acid residue region spanning residue R124 in FAS1-1 is capable of forming amyloid material in vitro upon the introduction of the R124C mutation, whereas the native peptide has a minor potential to form amyloid fibrils (Schmitt-Bernard et al. 2000). Consequently, the FAS1-1 domain may become amyloidogenic upon the introduction of the R124C mutation. This is supported by the observation that N-terminal fragments of TGFBIp accumulate in R124C corneas (Korvatska et al. 2000). In contrast, the amyloidogenic nature of FAS1-4 is suppressed in wild-type TGFBIp and is released only upon mutation-induced structural destabilization.

The thermodynamic stability of full-length TGFBIp mutants associated with GCD2 (R124H) and RBCD (R124L) is preserved when compared to wild-type TGFBIp (Runager et al. 2011). Consistent with R124 being surface exposed, X-ray crystallography does not reveal any changes in the monomeric structure of the R124H mutant compared to wild-type TGFBIp (Nielsen et al. 2021). Instead X-ray structures of R124H mutant demonstrate a new dimer interface in the

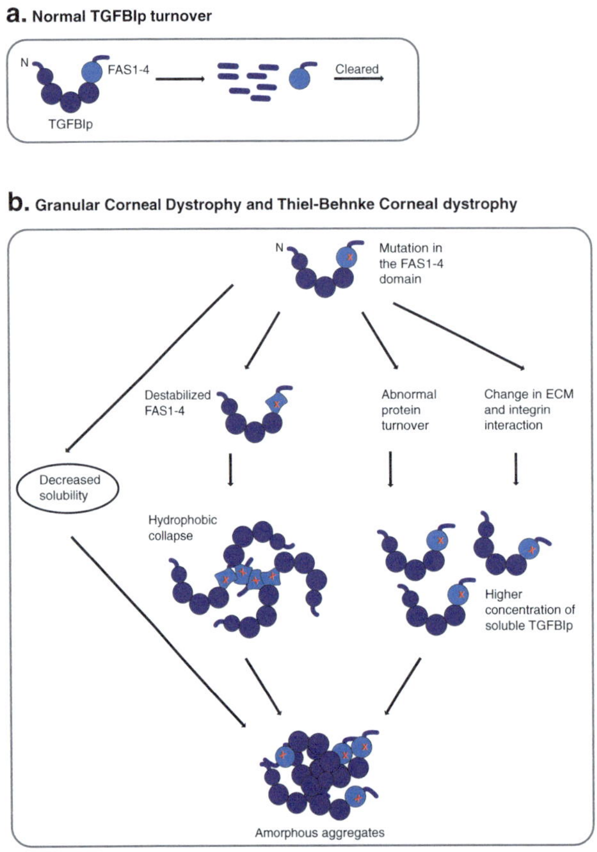

Fig. 9.4 Schematic representation of proposed mechanisms for *TGFBI*-linked corneal dystrophies associated with mutations in the FAS1–4 domain (red X). (**a**) In the normal human cornea, TGFBIp is extensively processed by the proteolytic machinery, leading to the generation of FAS1–4-containing products that are subsequently cleared. (**b**) In GCD and TBCD intact or near-intact TGFBIp accumulates. One possible mechanism for accumulation is reduced solubility of the folded monomeric state of the FAS1–4 domain. Another aggregation mechanism might be destabilization through hydrophobic collapse. Changes in TGFBIp interactions with the ECM or integrins or an abnormal turnover due to resistance to proteolytic degradation might lead to a higher concentration of soluble TGFBIp that could result in increased aggregation. (**c**) In LCD, fragments of TGFBIp accumulate. In general LCD-associated mutations in the FAS1–4 domain destabilize the structure of TGFBIp, causing the FAS1–4 domain to fibrillate directly and/or lead to a change in its proteolytic turnover by HtrA1 that liberates the amyloidogenic regions of the FAS1–4 domain. In addition, sulfated GAGs of the cornea might promote the fibrillation of TGFBIp. Adapted from Nielsen et al. (2020)

crystal packing, which is not present in wild-type TGFBIp structure or the structure of the LCD-associated mutant A546T. This interface, not confined to the R124H structure but also present for R555W, involves residue 124 of the first TGFBIp molecule and 555 in the second. The dimer interface is not permitted by the R124 and R555 residues of wild-type TGFBIp and may therefor play a central role in the aggregation exhibited by R124H and R555W in vivo (Fig. 9.4).

Concluding Remarks

Decades of research into the progression of TGFBIp-related protein aggregation in the cornea have highlighted phenotypic differences between GCD and LCD, driven by multiple underlying molecular mechanisms. Studies have shown that disease-causing mutations in the *TGFBI* gene lead to the production of TGFBIp with altered stability and solubility. This is believed to impact the turnover of corneal TGFBIp, resulting in the accumulation of intact protein due to halted turnover in GCD, or aberrant and increased turnover in the case of LCD progression (Fig. 9.4). This knowledge holds promise for the development of new therapeutic strategies to halt or delay TGFBIp aggregation in patients with GCD or LCD.

References

Andersen RB, Karring H, Moller-Pedersen T, Valnickova Z, Thogersen IB, Hedegaard CJ, Kristensen T, Klintworth GK, Enghild JJ. Purification and structural characterization of transforming growth factor beta induced protein (TGFBIp) from porcine and human corneas. Biochemistry. 2004;43:16374–84.

Andreasen M, Nielsen SB, Runager K, Christiansen G, Nielsen NC, Enghild JJ, Otzen DE. Polymorphic fibrillation of the destabilized fourth fasciclin-1 domain mutant A546T of the Transforming growth factor-beta-induced protein (TGFBIp) occurs through multiple pathways with different oligomeric intermediates. J Biol Chem. 2012;287:34730–42.

Ashburner M, Ball CA, Blake JA, Botstein D, Butler H, Cherry JM, Davis AP, Dolinski K, Dwight SS, Eppig JT, Harris MA, Hill DP, Issel-Tarver L, Kasarskis A, Lewis S, Matese JC, Richardson JE, Ringwald M, Rubin GM, Sherlock G. Gene ontology: tool for the unification of biology. The gene ontology consortium. Nat Genet. 2000;25:25–9.

Basaiawmoit RV, Oliveira CL, Runager K, Sorensen CS, Behrens MA, Jonsson BH, Kristensen T, Klintworth GK, Enghild JJ, Pedersen JS, Otzen DE. SAXS models of TGFBIp reveal a trimeric structure and show that the overall shape is not affected by the Arg124His mutation. J Mol Biol. 2011;408:503–13.

Bian YY, Song CX, Cheng K, Dong MM, Wang FJ, Huang JF, Sun DG, Wang LM, Ye ML, Zou HF. An enzyme assisted RP-RPLC approach for in-depth analysis of human liver phosphoproteome. J Proteomics. 2014;96:253–62.

Billings PC, Whitbeck JC, Adams CS, Abrams WR, Cohen AJ, Engelsberg BN, Howard PS, Rosenbloom J. The transforming growth factor-beta-inducible matrix protein (beta)ig-h3 interacts with fibronectin. J Biol Chem. 2002;277:28003–9.

Boote C, Hayes S, Jones S, Quantock AJ, Hocking PM, Inglehearn CF, Ali M, Meek KM. Collagen organization in the chicken cornea and structural alterations in the retinopathy, globe enlarged (rge) phenotype—an X-ray diffraction study. J Struct Biol. 2008;161:1–8.

Chen S, Mienaltowski MJ, Birk DE. Regulation of corneal stroma extracellular matrix assembly. Exp Eye Res. 2015;133:69–80.

Courtney DG, Poulsen ET, Kennedy S, Moore JE, Atkinson SD, Maurizi E, Nesbit MA, Moore CB, Enghild JJ. Protein composition of TGFBI-R124C— and TGFBI-R555W-associated aggregates suggests multiple mechanisms leading to lattice and granular corneal dystrophy. Invest Ophthalmol Vis Sci. 2015;56:4653–61.

Dyrlund TF, Poulsen ET, Scavenius C, Nikolajsen CL, Thogersen IB, Vorum H, Enghild JJ. Human cornea proteome: identification and quantitation of the proteins of the three main layers including epithelium, stroma, and endothelium. J Proteome Res. 2012;11:4231–9.

El Kochairi I, Letovanec I, Uffer S, Munier FL, Chaubert P, Schorderet DF. Systemic investigation of keratoepithelin deposits in TGFBI/BIGH3-related corneal dystrophy. Mol Vis. 2006;12:461–6.

Escribano J, Hernando N, Ghosh S, Crabb J, Coca-Prados M. cDNA from human ocular ciliary epithelium homologous to beta ig-h3 is preferentially expressed as an extracellular protein in the corneal epithelium. J Cell Physiol. 1994;160:511–21.

Galiacy SD, Froment C, Mouton-Barbosa E, Erraud A, Chaoui K, Desjardins L, Monsarrat B, Malecaze F, Burlet-Schiltz O. Deeper in the human cornea proteome using nanoLC-Orbitrap MS/MS: an improvement for future studies on cornea homeostasis and pathophysiology. J Proteomics. 2011;75:81–92.

Garcia-Castellanos R, Nielsen NS, Runager K, Thogersen IB, Lukassen MV, Poulsen ET, Goulas T, Enghild JJ, Gomis-Ruth FX. Structural and functional implications of human transforming growth factor beta-induced protein, TGFBIp, in corneal dystrophies. Structure. 2017;25:1740–1750.e1742.

Grau S, Baldi A, Bussani R, Tian X, Stefanescu R, Przybylski M, Richards P, Jones SA, Shridhar V, Clausen T, Ehrmann M. Implications of the serine protease HtrA1 in amyloid precursor protein processing. Proc Natl Acad Sci U S A. 2005;102:6021–6.

Han KE, Choi SI, Kim TI, Maeng YS, Stulting RD, Ji YW, Kim EK. Pathogenesis and treatments of TGFBI corneal dystrophies. Prog Retin Eye Res. 2016;50:67–88.

Hanssen E, Reinboth B, Gibson MA. Covalent and non-covalent interactions of betaig-h3 with collagen VI. Beta ig-h3 is covalently attached to the amino-terminal region of collagen VI in tissue microfibrils. J Biol Chem. 2003;278:24334–41.

Hashimoto K, Noshiro M, Ohno S, Kawamoto T, Satakeda H, Akagawa Y, Nakashima K, Okimura A, Ishida H, Okamoto T, Pan H, Shen M, Yan WQ, Kato Y. Characterization of a cartilage-derived 66-kDa protein (RGD-CAP/beta ig-h3) that binds to collagen. Biochim Biophys Acta. 1997;1355:303–14.

Juhl DW, Risor MW, Scavenius C, Rasmussen CB, Otzen D, Nielsen NC, Enghild JJ. Conservation of the amyloid interactome across diverse fibrillar structures. Sci Rep. 2019;9:3863.

Karring H, Thogersen IB, Klintworth GK, Moller-Pedersen T, Enghild JJ. A dataset of human cornea proteins identified by Peptide mass fingerprinting and tandem mass spectrometry. Mol Cell Proteomics. 2005;4:1406–8.

Karring H, Runager K, Valnickova Z, Thogersen IB, Moller-Pedersen T, Klintworth GK, Enghild JJ. Differential expression and processing of transforming growth factor beta induced protein (TGFBIp) in the normal human cornea during postnatal development and aging. Exp Eye Res. 2010;90:57–62.

Karring H, Runager K, Thogersen IB, Klintworth GK, Hojrup P, Enghild JJ. Composition and proteolytic processing of corneal deposits associated with mutations in the TGFBI gene. Exp Eye Res. 2012;96:163–70.

Karring H, Poulsen ET, Runager K, Thogersen IB, Klintworth GK, Hojrup P, Enghild JJ. Serine protease HtrA1 accumulates in corneal transforming growth factor beta induced protein (TGFBIp) amyloid deposits. Mol Vis. 2013;19:861–76.

Kim JE, Park RW, Choi JY, Bae YC, Kim KS, Joo CK, Kim IS. Molecular properties of wild-type and mutant beta IG-H3 proteins. Investig Ophthalmol Visual Sci. 2002;43:656–61.

Kim BY, Olzmann JA, Choi SI, Ahn SY, Kim TI, Cho HS, Suh H, Kim EK. Corneal dystrophy-associated R124H mutation disrupts TGFBI interaction with periostin and causes mislocalization to the lysosome. J Biol Chem. 2009;284:19580–91.

Kitamoto K, Taketani Y, Fujii W, Inamochi A, Toyono T, Miyai T, Yamagami S, Kuroda M, Usui T, Ouchi Y. Generation of mouse model of TGFBI-R124C corneal dystrophy using CRISPR/Cas9-mediated homology-directed repair. Sci Rep. 2020;10:2000.

Klintworth GK, Enghild JJ, Valnickova Z. Discovery of a novel protein (big-h3) in normal human cornea. Invest Ophthalmol Vis Sci. 1994;35:1938.

Klintworth GK, Valnickova Z, Enghild JJ. Accumulation of beta ig-h3 gene product in corneas with granular dystrophy. Am J Pathol. 1998;152:743–8.

Koldsø H, Andersen OJ, Nikolajsen CL, Scavenius C, Sørensen CS, Underhaug J, Runager K, Nielsen NC, Enghild JJ, Schiøtt B. Early events in the amyloid formation of the A546T mutant of transforming growth factor beta-induced protein in corneal dystrophies compared to the nonfibrillating R555W and R555Q mutants. Biochemistry. 2015;54:5546–56.

Kolozsvari L, Nogradi A, Hopp B, Bor Z. UV absorbance of the human cornea in the 240—to 400-nm range. Invest Ophthalmol Vis Sci. 2002;43:2165–8.

Korvatska E, Henry H, Mashima Y, Yamada M, Bachmann C, Munier FL, Schorderet DF. Amyloid and non-amyloid forms of 5q31-linked corneal dystrophy resulting from kerato-epithelin mutations at Arg-124 are associated with abnormal turnover of the protein. J Biol Chem. 2000;275:11465–9.

Liu Y, Huang H, Sun G, Alwadani S, Semba RD, Lutty GA, Yiu S, Edward DP. Gene expression profile of extracellular matrix and adhesion molecules in the human normal corneal stroma. Curr Eye Res. 2017;42:520–7.

Liu C, Miyajima T, Melangath G, Miyai T, Vasanth S, Deshpande N, Kumar V, Ong Tone S, Gupta R, Zhu S, Vojnovic D, Chen Y, Rogan EG, Mondal B, Zahid M, Jurkunas UV. Ultraviolet A light induces DNA damage and estrogen-DNA adducts in Fuchs endothelial corneal dystrophy causing females to be more affected. Proc Natl Acad Sci U S A. 2020;117:573–83.

Lukassen MV, Scavenius C, Thøgersen IB, Enghild JJ. Disulfide bond pattern of transforming growth factor β-induced protein. Biochemistry. 2016;55:5610–21.

Lukassen MV, Poulsen ET, Donaghy J, Mogensen EH, Christie KA, Roshanravan H, DeDioniso L, Nesbit MA, Moore T, Enghild JJ. Protein analysis of the TGFBI(R124H) mouse model gives insight into phenotype development of granular corneal dystrophy. Proteomics Clin Appl. 2020;14:e1900072.

Nielsen NS, Poulsen ET, Klintworth GK, Enghild JJ. Insight into the protein composition of immunoglobulin light chain deposits of eyelid, orbital and conjunctival amyloidosis. J Proteomics Bioinform. 2014;Suppl 8:002.

Nielsen NS, Juhl DW, Poulsen ET, Lukassen MV, Poulsen EC, Risor MW, Scavenius C, Enghild JJ. Mutation-induced deamidation of corneal dystrophy-related transforming growth factor beta-induced protein. Biochemistry. 2017;56:6470–80.

Nielsen NS, Poulsen ET, Lukassen MV, Chao Shern C, Mogensen EH, Weberskov CE, DeDionisio L, Schauser L, Moore TCB, Otzen DE, Hjortdal J, Enghild JJ. Biochemical mechanisms of aggregation in TGFBI-linked corneal dystrophies. Prog Retin Eye Res. 2020;77:100843.

Nielsen NS, Gadeberg TAF, Poulsen ET, Harwood SL, Weberskov CE, Pedersen JS, Andersen GR, Enghild JJ. Mutation-induced dimerization of transforming growth factor-β-induced protein may drive protein aggregation in granular corneal dystrophy. J Biol Chem. 2021;297:100858.

Pepys MB, Dyck RF, de Beer FC, Skinner M, Cohen AS. Binding of serum amyloid P-component (SAP) by amyloid fibrils. Clin Exp Immunol. 1979;38:284–93.

Poepsel S, Sprengel A, Sacca B, Kaschani F, Kaiser M, Gatsogiannis C, Raunser S, Clausen T, Ehrmann M. Determinants of amyloid fibril degradation by the PDZ protease HTRA1. Nat Chem Biol. 2015;11:862–9.

Poulsen ET, Dyrlund TF, Runager K, Scavenius C, Krogager TP, Hojrup P, Thogersen IB, Sanggaard KW, Vorum H, Hjortdal J, Enghild JJ. Proteomics of Fuchs' endothelial corneal dystrophy support that the extracellular matrix of Descemet's membrane is disordered. J Proteome Res. 2014a;13:4659–67.

Poulsen ET, Runager K, Risor MW, Dyrlund TF, Scavenius C, Karring H, Praetorius J, Vorum H, Otzen DE, Klintworth GK, Enghild JJ. Comparison of two phenotypically distinct lattice corneal dystrophies caused by mutations in the transforming growth factor beta induced (TGFBI) gene. Proteomics Clin Appl. 2014b;8:168–77.

Poulsen ET, Nielsen NS, Jensen MM, Nielsen E, Hjortdal J, Kim EK, Enghild JJ. LASIK surgery of granular corneal dystrophy type 2 patients leads to accumulation and differential proteolytic processing of transforming growth factor beta-induced protein (TGFBIp). Proteomics. 2016;16:539–43.

Poulsen ET, Pedersen KW, Marzeda AM, Enghild JJ. Serum amyloid P component (SAP) interactome in human plasma containing physiological calcium levels. Biochemistry. 2017;56:896–902.

Poulsen ET, Runager K, Nielsen NS, Lukassen MV, Thomsen K, Snider P, Simmons O, Vorum H, Conway SJ, Enghild JJ. Proteomic profiling of TGFBI-null mouse corneas reveals only minor changes in matrix composition supportive of TGFBI knockdown as therapy against TGFBI-linked corneal dystrophies. Febs J. 2018;285:101–14.

Poulsen ET, Nielsen NS, Scavenius C, Mogensen EH, Risor MW, Runager K, Lukassen MV, Rasmussen CB, Christiansen G, Richner M, Vorum H, Enghild JJ. The serine protease HtrA1 cleaves misfolded transforming growth factor beta-induced protein (TGFBIp) and induces amyloid formation. J Biol Chem. 2019;294:11817–28.

Puente XS, Sanchez LM, Overall CM, Lopez-Otin C. Human and mouse proteases: a comparative genomic approach. Nat Rev Genet. 2003;4:544–58.

Ramshini H, Parrini C, Relini A, Zampagni M, Mannini B, Pesce A, Saboury AA, Nemat-Gorgani M, Chiti F. Large proteins have a great tendency to aggregate but a low propensity to form amyloid fibrils. PLoS One. 2011;6:e16075.

Rawe IM, Zhan Q, Burrows R, Bennett K, Cintron C. Beta-ig. Molecular cloning and in situ hybridization in corneal tissues. Invest Ophthalmol Vis Sci. 1997;38:893–900.

Rawlings ND, Barrett AJ, Thomas PD, Huang X, Bateman A, Finn RD. The MEROPS database of proteolytic enzymes, their substrates and inhibitors in 2017 and a comparison with peptidases in the PANTHER database. Nucleic Acids Res. 2018;46:D624–d632.

Reinboth B, Thomas J, Hanssen E, Gibson MA. Beta ig-h3 interacts directly with biglycan and decorin, promotes collagen VI aggregation, and participates in ternary complexing with these macromolecules. J Biol Chem. 2006;281:7816–24.

Ricard-Blum S. The collagen family. Cold Spring Harb Perspect Biol. 2011;3:a004978.

Risør MW, Poulsen ET, Thomsen LR, Dyrlund TF, Nielsen TA, Nielsen NC, Sanggaard KW, Enghild JJ. The autolysis of human HtrA1 is governed by the redox state of its N-terminal domain. Biochemistry. 2014;53:3851–7.

Runager K, Garcia-Castellanos R, Valnickova Z, Kristensen T, Nielsen NC, Klintworth GK, Gomis-Ruth FX, Enghild JJ. Purification, crystallization and preliminary X-ray diffraction of wild-type and mutant recombinant human transforming growth factor beta-induced protein (TGFBIp). Acta Crystallogr Sect F Struct Biol Cryst Commun. 2009;65:299–303.

Runager K, Basaiawmoit RV, Deva T, Andreasen M, Valnickova Z, Sorensen CS, Karring H, Thogersen IB, Christiansen G, Underhaug J, Kristensen T, Nielsen NC, Klintworth GK, Otzen DE, Enghild JJ. Human phenotypically distinct TGFBI corneal dystrophies are linked to the stability of the fourth FAS1 domain of TGFBIp. J Biol Chem. 2011;286:4951–8.

Runager K, Klintworth GK, Karring H, Enghild JJ. The insoluble TGFBIp fraction of the cornea is covalently linked via a disulfide bond to type XII collagen. Biochemistry. 2013;52:2821–7.

Runyon ST, Zhang Y, Appleton BA, Sazinsky SL, Wu P, Pan B, Wiesmann C, Skelton NJ, Sidhu SS. Structural and functional analysis of the PDZ domains of human HtrA1 and HtrA3. Protein Sci. 2007;16:2454–71.

Schmitt-Bernard CF, Chavanieu A, Derancourt J, Arnaud B, Demaille JG, Calas B, Argiles A. In vitro creation of amyloid fibrils from native and Arg124Cys mutated betaIGH3((110–131)) peptides, and its relevance for lattice corneal amyloid dystrophy type I. Biochem Biophys Res Commun. 2000;273:649–53.

Semba RD, Enghild JJ, Venkatraman V, Dyrlund TF, Van Eyk JE. The human eye proteome project: perspectives on an emerging proteome. Proteomics. 2013;13:2500–11.

Sørensen CS, Runager K, Scavenius C, Jensen MM, Nielsen NS, Christiansen G, Petersen SV, Karring H, Sanggaard KW, Enghild JJ. Fibril core of transforming growth factor beta-induced protein (TGFBIp) facilitates aggregation of corneal TGFBIp. Biochemistry. 2015;54:2943–56.

Srivastava OP, Srivastava K. cAMP-dependent phosphorylation of betaig-h3 protein in human corneal endothelial cells. Curr Eye Res. 1999;19:348–57.

Stagos D, Chen Y, Cantore M, Jester JV, Vasiliou V. Corneal aldehyde dehydrogenases: multiple functions and novel nuclear localization. Brain Res Bull. 2010;81:211–8.

Stenvang M, Schafer NP, Malmos KG, Perez AW, Niembro O, Sormanni P, Basaiawmoit RV, Christiansen G, Andreasen M, Otzen DE. Corneal dystrophy mutations drive pathogenesis by targeting TGFBIp stability and solubility in a latent amyloid-forming domain. J Mol Biol. 2018;430:1116–40.

Takacs L, Csutak A, Balazs E, Modis L Jr, Berta A. Expression of betaig-h3 is lower than normal in keratoconus corneas but increases with scarring. Cornea. 1999;18:599–605.

Tamura Y, Konomi H, Sawada H, Takashima S, Nakajima A. Tissue distribution of type VIII collagen in human adult and fetal eyes. Invest Ophthalmol Vis Sci. 1991;32:2636–44.

Tennstaedt A, Popsel S, Truebestein L, Hauske P, Brockmann A, Schmidt N, Irle I, Sacca B, Niemeyer CM, Brandt R, Ksiezak-Reding H, Tirniceriu AL, Egensperger R, Baldi A, Dehmelt L, Kaiser M, Huber R, Clausen T, Ehrmann M. Human high temperature requirement serine protease A1 (HTRA1) degrades tau protein aggregates. J Biol Chem. 2012;287:20931–41.

Truebestein L, Tennstaedt A, Mönig T, Krojer T, Canellas F, Kaiser M, Clausen T, Ehrmann M. Substrate-induced remodeling of the active site regulates human HTRA1 activity. Nat Struct Mol Biol. 2011;18:386–8.

Underhaug J, Koldsø H, Runager K, Nielsen JT, Sørensen CS, Kristensen T, Otzen DE, Karring H, Malmendal A, Schiøtt B, Enghild JJ, Nielsen NC. Mutation in transforming growth factor beta induced protein associated with granular corneal dystrophy type 1 reduces the proteolytic susceptibility through local structural stabilization. Biochim Biophys Acta. 2013;1834:2812–22.

Venkatraman A, Dutta B, Murugan E, Piliang H, Lakshminaryanan R, Sook Yee AC, Pervushin KV, Sze SK, Mehta JS. Proteomic analysis of amyloid corneal aggregates from TGFBI-H626R lattice corneal dystrophy patient implicates serine-protease HTRA1 in mutation-specific pathogenesis of TGFBIp. J Proteome Res. 2017;16:2899–913.

Vierkotten S, Muether PS, Fauser S. Overexpression of HTRA1 leads to ultrastructural changes in the elastic layer of Bruch's membrane via cleavage of extracellular matrix components. PLoS One. 2011;6:e22959.

Yamazoe K, Yoshida S, Yasuda M, Hatou S, Inagaki E, Ogawa Y, Tsubota K, Shimmura S. Development of a transgenic mouse with R124H human TGFBI mutation associated with granular corneal dystrophy type 2. PLoS One. 2015;10:e0133397.

Zurawa-Janicka D, Skorko-Glonek J, Lipinska B. HtrA proteins as targets in therapy of cancer and other diseases. Exp Opin Ther Targets. 2010;14:665–79.

Genetics and Gene Therapy of the *TGFBI* Corneal Dystrophies

10

Amanda M. Kowalczyk, M. Andrew Nesbit, Andrew McLain, and C. B. Tara Moore

TGFBI Corneal Dystrophies

Corneal Dystrophies

Corneal dystrophies were first described in 1890 by Arthur Groenouw as *noduli corneae* in patients with granular and macular corneal dystrophies (Wittebol-Post et al. 1987). In 1899, Biber published findings on lattice corneal dystrophy (LCD) (Biber 1890). The phenotype and familial distribution were yet to be investigated. In 1917, Reis published findings on an unknown superficial dystrophy, while Thiel-Behnke investigated a honeycomb-patterned variant of superficial dystrophy. The literature at that time remained ambiguous as to whether these dystrophies were distinct (Küchle et al. 1995). In 1967, Klintworth determined that the lattice pattern in LCD was similar to that of amyloid fibrils and that they could be stained with congo red and thioflavin T. This study provided the first evidence that amyloid deposits were associated with corneal dystrophies and suggested that collagen degradation was a central characteristic of the disease (Klintworth 1967).

In 1949, Bücklers published the first classification of corneal dystrophies based on anatomical location, classifying corneal dystrophies according to the layer(s) of cornea that were affected. Classification was purely based on physical appearance without consideration of inheritance patterns, severity, histopathologic, or biochemical characteristics (Bücklers 1949). In 1988, Folberg described a combined lattice-granular dystrophy in four patients and traced family origins to Italian province of Avellino after which it became known as Avellino corneal dystrophy (ACD). ACD was later reclassified as granular corneal dystrophy type 2 (GCD2) (Folberg et al. 1988).

The transforming growth factor beta-induced (*TGFBI*) gene was characterized in 1992 (Skonier et al. 1992). Subsequently, the expression of TGFBI was reported in the cornea and verified at the protein level, showing increased protein expression in pathological conditions like granular dystrophy (Klintworth 1999).

In 1994, Stone et al. described three autosomal dominant corneal dystrophies: lattice corneal dystrophy type I, granular dystrophy, and

A. M. Kowalczyk
Integrated Diagnostic Laboratory, Ulster University, Belfast, Northern Ireland, UK

M. A. Nesbit · A. McLain
Integrated Diagnostic Laboratory, Ulster University, Belfast, Northern Ireland, UK

Biomedical Sciences Research Institute, Ulster University, Coleraine, Northern Ireland, UK

C. B. T. Moore (✉)
Integrated Diagnostic Laboratory, Ulster University, Belfast, Northern Ireland, UK

Biomedical Sciences Research Institute, Ulster University, Coleraine, Northern Ireland, UK

Avellino, Menlo Park, CA, USA
e-mail: tara.moore@ulster.ac.uk

E. K. Kim, R. D. Stulting (eds.), *TGFBI-related Corneal Dystrophies*,
https://doi.org/10.1007/978-981-96-0131-8_10

Avellino dystrophy, mapping them to the long arm of chromosome 5q (Stone et al. 1994). Korvatska et al. reported missense mutations in the *TGFBI* gene, formerly known as *BIGH3* gene, on two hotspots at chromosome 5q31: R124 and R555 (Korvatska et al. 1998). Haplotype analysis indicated that the mutations did not originate from a common ancestor since they occurred independently in several ethnic groups (Korvatska et al. 1998). Subsequent studies show that five mutations at these two codons account for 75% of the reported cases (Chao-Shern et al. 2019) (Fig. 10.1).

Mutations in the *TGFBI* gene cause a spectrum of distinct epithelial and stromal corneal dystrophies resulting in corneal amyloid and non-amyloid deposits. Some dystrophies have very distinct patterns that make the diagnosis relatively easy. However, some may have similar symptoms and clinical findings. For example, lattice corneal dystrophy and granular corneal dystrophy are very easily distinguished from one another (Fig. 10.1). Lattice corneal dystrophy has refractile amyloid lines, while granular corneal dystrophy has well-defined deposits resembling crushed bread crumbs (Klintworth 2009; Weiss et al. 2015). RBCD and TBCD both have irregular ring-shaped patterns of spots and lines (Fig. 10.1) (Klintworth 2009; Weiss et al. 2015).

It is difficult to differentiate between RBCD and TBCD without histology or genetic testing (Klintworth 2009). The histologic appearance of RBCD is that of "rod-shape" deposits in the corneal epithelium and stroma, while TBCD is characterized by curly fibers that stain for keratoepithelin (23). Compared to the histological

Corneal Dystrophy Name	Gene	Chromosome Location	Mutation	Clinical Phenotype	Image
GCD1	*TGFBI*	5q31.1	p.R555W	bilateral deposits; snowflake-like appearance with well-defined granules	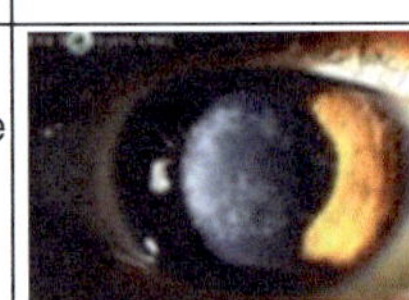
TBCD	*TGFBI*	5q31.1	p.R555Q	single large honeycomb-like opacity	
RBCD	*TGFBI*	5q31.1	p.R124L; G623D	cloud of superficial gray-white opacities	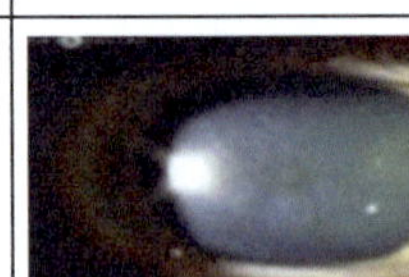
LCD	*TGFBI*	5q31.1	p.R124C	thin branching retractile lines	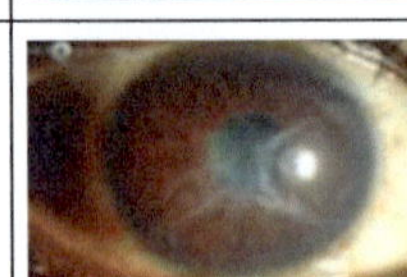
GCD2	*TGFBI*	5q31.1	p.R124H	bilateral progression of subtle superficial stromal whitish dots to ring/snowflake shaped gray-white granular deposits in the anterior stroma	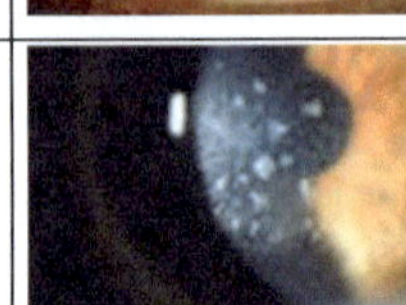

Fig. 10.1 A summary of the different corneal dystrophies associated with mutations in the *TGFBI* gene including the location of the mutation and a summary of the clinical phenotype

approach where cornea tissue is required, genetic testing provides a noninvasive method by which RBCD and TBCD can be readily differentiated by analysis of buccal swabs.

To date, 17 coding exons and 73 mutations associated with corneal dystrophies have been reported and mapped to their respective exons (Table 10.1).

Genetics

Hotspots at codons 124 in exon 4 and 555 in exon 12 are responsible for the most common *TGFBI* corneal dystrophies. Eight missense *TGFBI* mutations, one deletion, and one insertion are disease-causing mutations of LCD, intermediate type I/IIIA. Nineteen missense mutations and three deletions are the causes of other types of lattice type corneal dystrophy. Most of the *TGFBI* corneal dystrophy mutations reside on exons 4, 11, 12, and 14 (Han et al. 2016).

The worldwide prevalence of *TGFBI*-related corneal dystrophies is unknown; however, estimates have been published for certain populations. The estimated prevalence for GCD2 in Korea is 1:870 (Lee et al. 2010); for the five most common *TGFBI* corneal dystrophies, a prevalence of 1:476 was reported in China (Song et al. 2017); in the USA according to insurance claims data, a prevalence of 1:1115 for all corneal dystrophies was reported (Musch et al. 2011). Over 1600 reported cases were reviewed to explore the *TGFBI* mutation distribution in specific populations and geographical regions with no overall differences found (12). *TGFBI* mutation cases have been reported from all over the world, except for regions with limited research capacity or language barriers to publication (Chao-Shern et al. 2019). Of 1633 cases reviewed, R124C LCD1 accounted for 372 cases (23% of the total), R124H GCD2 for 325 cases (20%), R124L RBCD for 110 cases (7%), R555Q TBCD for 75 cases (5%), R555W GCD1 for 338 cases (21%), and the remaining 65 mutations accounted for 24% of cases reported worldwide. Therefore, the five most common mutations only account for about 76% of the *TGFBI*-related mutations.

We identified another six mutations that accounted for a portion of the remaining reported cases. R124S, a subtype of GCD, accounted for 18 cases, which was 1.1% of the total; A546D, a variant of LCD, accounted for 48 cases (2.9%); H572R, a subtype of LCD, accounted for 34 cases (2.1%); G623D accounted for 26 cases (1.6%); H626R, LCD subtype, accounted for 117 cases (7.2%); and M502V, a variant of CD, accounted for 4 cases (0.25%). We determined that the next six most common mutations account for about 15% of the published cases. To increase the detection rate of *TGFBI* corneal dystrophy by the five SNP test which detects the five most prevalent mutations at the R124 and R555 hotspots (R124H, R124C, R124L, R555W, and R555Q), we developed an additional six point mutation real-time PCR assay (Chao-Shern et al. 2019). The new 11 (5 plus 6) mutation panel increases the detection rate to over 90%. The majority of the remaining mutations are found in individuals or single families, so these will not be detectable by such a real-time PCR assay method. Instead, advances in DNA sequencing technology, and inclusion of *TGFBI* amplicons in targeted resequencing panels, have enabled detection of these very rare mutations where the corneal dystrophy is clearly present clinically (AvaGen, Avellino).

In addition to a clinical examination and exclusion of family history, and importantly in the absence of clinical pathology, genetic prescreening for corneal dystrophy mutations in pre-refractive surgery candidates is key to avoiding laser-induced onset or exacerbation of symptoms and corneal pathology.

Global Prevalence

To date, over 70 *TGFBI* mutations are listed in the human gene mutation database (HGMD) as causing a spectrum of different epithelial-stromal corneal dystrophies with corneal amyloid and non-amyloid deposits, including granular corneal dystrophy type 1 (GCD1) and type 2 (GCD2), lattice corneal dystrophy (LCD), Reis-Bücklers corneal dystrophy (RBCD), and Thiel-Behnke

Table 10.1 Human gene mutation database (HGMD) (Stenson et al. 2017) (Qiagen, Hilden, Germany) listed 70 *TGFBI* mutations, small deletion/insertion, and small indels. The genomic coordinates, dbSNP ID (https://www.ncbi.nlm.nih.gov/snp/), exons, type of mutations, amino acid change, protein change, reported phenotypes, first reference, and reported mode of inheritance are described

Genomic coordinates	rs#	Exon	Mutations	Amino acid change	Nucleotide change	Protein change	HGMD reported phenotype	First reference (Credit: HGMD)	First reported mode of inheritance
chr5:135382062	rs757933370	4	Missense	Val113Ile	c.337G > A	p.V113I	Corneal dystrophy, granular	Zenteno (2006) Mol Vis 12, 331	AD
chr5:135382092	rs541270955	4	Missense	Asp123His	c.367G > C	p.D123H	Corneal dystrophy, granular	Ha (2003) Jpn J Ophthalmol 47, 246	AD
chr5:135382095	rs121909210	4	Missense	Arg124Cys	c.370C > T	p.R124C	Corneal dystrophy, lattice type I	Munier (1997) Nat Genet 15, 247	AD
chr5:135382096	rs121909211	4	Missense	Arg124His	c.371G > A	p.R124H	Corneal dystrophy, Avellino	Munier (1997) Nat Genet 15, 247	AD
chr5:135382096	rs121909211	4	Missense	Arg124Leu	c.371G > T	p.R124L	Corneal dystrophy, Reis-Bücklers	Okada (1998) Am J Ophthalmol 126, 535	AD
chr5:135382095	rs121909210	4	Missense	Arg124Ser	c.370C > A	p.R124S	Corneal dystrophy, granular type I	Stewart (1999) Hum Mutat 14, 126	AD
chr5:135382118	rs199604416	4	Missense	Glu131Asp	c.393G > T	p.E131D	Corneal dystrophy, Schnyder	Foja (2016) Int Ophthalmol 36, 867	AD
chr5:135382615	rs886059924	5	Missense	Arg179Term	c.535C > T	p.R179*	Corneal dystrophy, granular type II	Song et al. (2015) Ann Lab Med 35, 336	Not reported
chr5:135391444	rs10057190	11	Missense	Arg496Trp	c.1486C > T	p.R496W	Corneal dystrophy, lattice type	Kawasaki (2011) Br J Ophthalmol 95, 150	Not reported
chr5:135391459	rs121909212	11	Missense	Pro501Thr	c.1501C > A	p.P501T	Corneal dystrophy, lattice type IIIA	Yamamoto (1998) Am J Hum Genet 62, 719	AD
chr5:135391462	rs188677757	11	Missense	Met502Val	c.1504A > G	p.M502V	Corneal dystrophy	Zenteno (2009) Exp Eye Res 89, 172	AD
chr5:135391472	No dbSNP ID	11	Missense	Val505Asp	c.1514T > A	p.V505D	Corneal dystrophy, lattice type I	Tian (2005) Jpn J Ophthalmol 49, 84	AD
chr5:135391484	rs121909216	11	Missense	Leu509Arg	c.1526T > G	p.L509R	Corneal dystrophy, epithelial basement membrane	Boutboul (2006) Hum Mutat 27, 553	AD

Genomic coordinates	rs#	Exon	Mutations	Amino acid change	Nucleotide change	Protein change	HGMD reported phenotype	First reference (Credit: HGMD)	First reported mode of inheritance
chr5:135391484	No dbSNP ID	11	Missense	Leu509Pro	c.1526T > C	p.L509P	Corneal dystrophy, Reis-Bücklers	Gruenauer-Kloevekorn (2009) Br J Ophthalmol 93, 932	AD
chr5:135391499	No dbSNP ID	11	Missense	Arg514Pro	c.1541G > C	p.R514P	Corneal dystrophy, lattice type	Zhong (2010) Mol Vis 16, 224	AD
chr5:135391503	No dbSNP ID	11	Missense	Phe515Leu	c.1545T > A	p.F515L	Corneal dystrophy, lattice type	Zhong (2010) Mol Vis 16, 224	AD
chr5:135392354	No dbSNP ID	12	Missense	Ser516Arg	c.1548C > G	p.S516R	Corneal dystrophy	Paliwal (2010) Mol Vis 16, 1429	AD
chr5:135392359	No dbSNP ID	12	Missense	Leu518Arg	c.1553T > G	p.L518R	Corneal dystrophy, lattice intermediate type I/IIIA	Munier (2002) Invest Ophthalmol Vis Sci 43, 949	AD
chr5:135392359	No dbSNP ID	12	Missense	Leu518Pro	c.1553T > C	p.L518P	Corneal dystrophy, lattice intermediate type I/IIIA	Endo (1999) Am J Ophthalmol 128, 104	AD
chr5:135392371	No dbSNP ID	12	Missense	Ile522Asn	c.1565T > A	p.I522N	Corneal dystrophy, lattice type I	Zhang (2009) Mol Vis 15, 2498	AD
chr5:135392386	rs1050842080	12	Missense	Leu527Arg	c.1580T > G	p.L527R	Corneal dystrophy, lattice type, with deep deposits	Fujiki et al. (1998) Hum Genet 103, 286	AD
chr5:135392419	No dbSNP ID	12	Missense	Thr538Arg	c.1613C > G	p.T538R	Corneal dystrophy, lattice intermediate type I/IIIA	Munier (2002) Invest Ophthalmol Vis Sci 43, 949	AD
chr5:135392418	No dbSNP ID	12	Missense	Thr538Pro	c.1612A > C	p.T538P	Corneal dystrophy, lattice type	Yu (2006) J Genet 85, 73	AD
chr5:135392422	rs1382893670	12	Missense	Val539Asp	c.1616T > A	p.V539D	Corneal dystrophy, lattice type	Chakravarthi (2005) Invest Ophthalmol Vis Sci 46, 121	AD
chr5:135392425	rs121909214	12	Missense	Phe540Ser	c.1619T > C	p.F540S	Corneal dystrophy, lattice type	Stix (2005) Invest Ophthalmol Vis Sci 46, 1133	AD
chr5:135392431	No dbSNP ID	12	Missense	Pro542Arg	c.1625C > G	p.P542R	Corneal dystrophy, lattice type	Cho (2012) Mol Vis 18, 2012	AD

(continued)

Table 10.1 (continued)

Genomic coordinates	rs#	Exon	Mutations	Amino acid change	Nucleotide change	Protein change	HGMD reported phenotype	First reference (Credit: HGMD)	First reported mode of inheritance
chr5:135392437	rs777288957	12	Missense	Asn544Ser	c.1631A > G	p.N544S	Corneal dystrophy, lattice intermediate type I/IIIA	Mashima (2000) Am J Ophthalmol 130, 516	AD
chr5:135392443	rs267607109	12	Missense	Ala546Asp	c.1637C > A	p.A546D	Corneal dystrophy, lattice type	Aldave (2004) Am J Ophthalmol 138, 772	AD
chr5:135392442	No dbSNP ID	12	Missense	Ala546Thr	c.1636G > A	p.A546T	Corneal dystrophy, lattice type IIIA	Dighiero (2000) Am J Ophthalmol 129, 248	AD
chr5:135392446	No dbSNP ID	12	Missense	Phe547Cys	c.1640T > G	p.F547C	Corneal dystrophy, granular	Foja (2016) Int Ophthalmol 36, 867	Not reported
chr5:135392446	No dbSNP ID	12	Missense	Phe547Ser	c.1640T > C	p.F547S	Corneal dystrophy, lattice type	Takács (2007) Mol Vis 13, 1976	AD
chr5:135392449	No dbSNP ID	12	Missense	Arg548Pro	c.1643G > C	p.R548P	Corneal dystrophy, lattice type	Chae (2016) Clin Genet 89, 678	Not reported
chr5:135392455	No dbSNP ID	12	Missense	Leu550Pro	c.1649T > C	p.L550P	Corneal dystrophy, granular	Zenteno (2009) Exp Eye Res 89, 172	AD
chr5:135392458	rs267607110	12	Missense	Pro551Gln	c.1652C > A	p.P551Q	Corneal dystrophy, lattice type	Aldave (2004) Am J Ophthalmol 138, 772	AD
chr5:135392470	rs121909209	12	Missense	Arg555Gln	c.1664G > A	p.R555Q	Corneal dystrophy, Thiel-Behnke	Munier (1997) Nat Genet 15, 247	AD
chr5:135392469	rs121909208	12	Missense	Arg555Trp	c.1663C > T	p.R555W	Corneal dystrophy, granular type I	Munier (1997) Nat Genet 15, 247	AD
chr5:135392479	No dbSNP ID	12	Missense	Leu558Arg	c.1673T > G	p.L558R	Corneal dystrophy, lattice type	Dudakova (2016) Ophthalmic Genet 37, 473	Not reported
chr5:135392479	No dbSNP ID	12	Missense	Leu558Pro	c.1673T > C	p.L558P	Corneal dystrophy, lattice type	Livshits (2008) Hum Genet 124, 296	Not reported
chr5:135392481	No dbSNP ID	12	Missense	Leu559Val	c.1675T > G	p.L559V	Corneal dystrophy	Paliwal (2010) Mol Vis 16, 1429	AD
chr5:135394794	No dbSNP ID	12	Missense	Leu565Pro	c.1694T > C	p.L565P	Corneal dystrophy, lattice type	Ołdak (2014) Cornea 33, 294	Not reported

Genomic coordinates	rs#	Exon	Mutations	Amino acid change	Nucleotide change	Protein change	HGMD reported phenotype	First reference (Credit: HGMD)	First reported mode of inheritance
chr5:135394806	No dbSNP ID	13	Missense	Leu569Arg	c.1706T > G	p.L569R	Corneal dystrophy, lattice type I	Warren (2003) Am J Ophthalmol 136, 872	AD
chr5:135394806	No dbSNP ID	13	Missense	Leu569Gln	c.1706T > A	p.L569Q	Corneal dystrophy, lattice type I	Song et al. (2015) Ann Lab Med 35, 336	Not reported
chr5:135394815	No dbSNP ID	13	Missense	His572Arg	c.1715A > G	p.H572R	Corneal dystrophy, lattice type I	Atchaneeyasakul (2006) Jpn J Ophthalmol 50, 403	AD
chr5:135394881	No dbSNP ID	13	Missense	Gly594Val	c.1781G > T	p.G594V	Corneal dystrophy, lattice type	Chakravarthi (2005) Invest Ophthalmol Vis Sci 46, 121	AD
chr5:135396557	No dbSNP ID	14	Missense	Val613Gly	c.1838T > G	p.V613G	Corneal dystrophy, lattice type	Niel-Butschi (2011) Mol Vis 17, 1192	AD
chr5:135396575	No dbSNP ID	14	Missense	Met619Lys	c.1856T > A	p.M619K	Corneal dystrophy, combined granular-lattice type, variant of	Aldave (2008) Arch Ophthalmol 126, 371	AD
chr5:135396578	No dbSNP ID	14	Missense	Ala620Asp	c.1859C > A	p.A620D	Corneal dystrophy, lattice type	Lakshminarayanan (2011) Br J Ophthalmol 95, 1457	Not reported
chr5:135396577	No dbSNP ID	14	Missense	Ala620Pro	c.1858G > C	p.A620P	Corneal dystrophy, lattice type IIIA	Jung (2014) Cornea 33, 1324	Not reported
chr5:135396580	No dbSNP ID	14	Missense	Thr621Pro	c.1861A > C	p.T621P	Corneal dystrophy, lattice type I	Song et al. (2015) Ann Lab Med 35, 336	Not reported
chr5:135396583	No dbSNP ID	14	Missense	Asn622His	c.1864A > C	p.N622H	Corneal dystrophy, lattice intermediate type I/IIIA	Stewart (1999) Ophthalmology 106, 964	AD
chr5:135396585	No dbSNP ID	14	Missense	Asn622Lys	c.1866T > A	p.N622K	Corneal dystrophy, lattice type IIIA	Munier (2002) Invest Ophthalmol Vis Sci 43, 949	AD

(continued)

Table 10.1 (continued)

Genomic coordinates	rs#	Exon	Mutations	Amino acid change	Nucleotide change	Protein change	HGMD reported phenotype	First reference (Credit: HGMD)	First reported mode of inheritance
chr5:135396586	No dbSNP ID	14	Missense	Gly623Arg	c.1867G > C	p.G623R	Corneal dystrophy, map-like	Gruenauer-Kloevekorn (2009) Graefes Arch Clin Exp Ophthalmol 247, 93	AD
chr5:135396587	rs121909215	14	Missense	Gly623Asp	c.1868G > A	p.G623D	Corneal dystrophy, lattice intermediate type I/IIIA	Afshari (2001) Arch Ophthalmol 119, 16	Not reported
chr5:135396589	rs200201691	14	Missense	Val624Met	c.1870G > A	p.V624M	Corneal dystrophy, lattice type	Kannabiran (2006) Hum Mutat 27, 615	AD
chr5:135396593	No dbSNP ID	14	Missense	Val625Asp	c.1874T > A	p.V625D	Corneal dystrophy, lattice type	Tian (2007) Am J Ophthalmol 144, 473	AD
chr5:135396596	rs1052006472	14	Missense	His626Arg	c.1877A > G	p.H626R	Corneal dystrophy, lattice intermediate type I/IIIA	Stewart (1999) Ophthalmology 106, 964	AD
chr5:135396596	No dbSNP ID	14	Missense	His626Pro	c.1877A > C	p.H626P	Corneal dystrophy, lattice intermediate type I/IIIA	Munier (2002) Invest Ophthalmol Vis Sci 43, 949	AD
chr5:135396611	No dbSNP ID	14	Missense	Val631Asp	c.1892T > A	p.V631D	Corneal dystrophy, lattice type, with deep deposits	Munier (2002) Invest Ophthalmol Vis Sci 43, 949	AD
chr5:135398363	rs121909217	16	Missense	Arg666Ser	c.1998G > C	p.R666S	Corneal dystrophy, epithelial basement membrane	Boutboul (2006) Hum Mutat 27, 553	AD
chr5:135382035 – 135382036	No dbSNP ID	4	Small deletions CCCTA^102CCACTct-CAAACCTTTA	CCCTA^102-CCACTct-CAAACCTTTA	c.310_311delTC	p.(Ser104-Lysfs*27)	Corneal dystrophy, granular type I	Pang (2002) Hum Mutat 19, 189	AD
chr5:135392424 – 135392426	No dbSNP ID	12	Small deletions CTACACA^539-GTCtttGCT-CCCACAA	CTACACA^539-GTCtttGCT-CCCACAA	c.1618_1620delTTT	p.(Phe540del)	Corneal dystrophy, lattice intermediate type I/IIIA	Rozzo (1998) Hum Mutat 12, 215	AD
chr5:135394814 – 135394816	No dbSNP ID	13	Small deletions CCTGAAA^571-TACcacATT-GGTGATG	CCTGAAA^571-TACcacATT-GGTGATG	c.1714_1716delCAC	p.(His572del)	Corneal dystrophy, lattice type	Aldave (2006) Mol Vis 12, 142	Not reported

Genomic coordinates	rs#	Exon	Mutations	Amino acid change	Nucleotide change	Protein change	HGMD reported phenotype	First reference (Credit: HGMD)	First reported mode of inheritance
chr5:135396557 – 135396568	No dbSNP ID	14	Small deletions AAGGAG^^{612}-CCTGttgccgagcctg-ACATCATGGC	AAGGAG^^{612}CC-TGttgccgagcctg-ACATCATGGC	c.1838_1849del12	p.(Val613_Pro616del)	Corneal dystrophy, lattice type	Yang (2010) Mol Vis 16, 1186	AD
chr5:135396589 – 135396594	No dbSNP ID	14	Small deletions CACAAAT^^{623}-GGCgtggtc-CATGTCATCA	CACAAAT^^{623}-GGCgtggtc-CATGTCATCA	c.1870_1875-delGTGGTC	p.(Val624_Val625del)	Corneal dystrophy, lattice type	Chakravarthi (2005) Invest Ophthalmol Vis Sci 46, 121	AD
chr5:135396598 – 135396598	No dbSNP ID	14	Small deletions CGTGGTC^^{626}-CATgTCATCACCAA	CGTGGTC^^{626}-CATgTCA-TCACCAA	c.1879delG	p.(Val627-Serfs*44)	Corneal dystrophy, lattice type IIIA	Munier (2002) Invest Ophthalmol Vis Sci 43, 949	AD
chr5:135396613 – 135396614	No dbSNP ID	14	Small insertions TGTCATC^^{629}-ACCaatgttccc-AATGTTCTGC	TGTCATC^^{629}-ACCaatgttccc-AATGTTCTGC	c.1886_1894dup-CCAATGTTC	N/A	Corneal dystrophy, lattice intermediate type I/IIIA	Schmitt-Bernard (2000) Invest Ophthalmol Vis Sci 41, 1302	AD
chr5:135382096 – 135382103	No dbSNP ID	4	Small indels TACACG^^{123}-GACCgcacggag-AAGCTGAGGC Ins:tc	TACACG^^{123}-GACCgcacggag-AAGCTGAGGC Ins:tc	c.371_378delGC-ACGGAGinsTC	N/A	Corneal dystrophy, granular type I	Dighiero (2000) Arch Ophthalmol 118, 814	AD

corneal dystrophy (TBCD) (Evans et al. 2016; Korvatska et al. 2000). Different *TGFBI* mutations can cause specific corneal dystrophies, and a genotype-phenotype correlation has been demonstrated at two mutation hotspots, R124 and R555 (Evans et al. 2016). Overall, 75% of the *TGFBI* mutations reported in the over 1600 cases consisted of 1 of the 5 mutations currently detected by the available genetic test. While reports of novel *TGFBI* mutations are likely to be published, the most common *TGFBI* mutations, found at codons R124 and R555, are conversely underreported. Therefore, it is difficult to obtain an accurate estimation of the true worldwide detection rate of *TGFBI* dystrophies by reviewing the literature.

Global Available Genetic Test Data Analysis

Since 2008, more than 600,000 samples worldwide have been tested by the available genetic tests. Most of the samples were from Korea and Japan, where genetic testing is commonly used to screen patients prior to refractive surgery. An analysis of the global testing data demonstrated that the detection rate in Korea is approximately 15 in 10,000 people, which closely matches the reported prevalence of 1 in 870 people (Lee et al. 2010). The detection rate for *TGFBI* mutations in Japan (3 in 10,000) is lower than that in Korea. In Korea, the test is administered as a general screening for all refractive surgery candidates, whereas in Japan, patients are first subjected to a rigorous clinical examination and only samples from patients who have no detected corneal abnormalities are submitted for genetic testing.

Clinics and hospitals in Korea and Japan use the genetic test for screening purposes, as it is part of the practice guidelines for refractive surgery. In the USA, some clinics and hospitals use the test for screening during the preoperative examination for refractive corneal surgery, while others use it to confirm or exclude *TGFBI* mutations if the surgeon notices any corneal findings that are suggestive of corneal dystrophies. European clinics also utilize the test primarily in this manner.

Assessment of an Expanded Panel with Six Additional Mutations

In 2016, Moorfield's Corneal Dystrophy Study was carried out to conduct Sanger sequencing on the entire *TGFBI* gene (Evans et al. 2016). This study provided a dataset that allowed the identification of six new mutations sites, enhancing the detection rate in any given population. In brief, the study consisted of 91 unrelated *TGFBI* corneal dystrophy cases, 68 of which had a diagnosis of epithelial-stromal *TGFBI*-associated dystrophy (RBCD, TBCD, LCD, and GCD) and 23 had a diagnosis of bilateral EBMD (Evans et al. 2016).

Treatment of Genetic Corneal Diseases

Currently, corneal dystrophies are treated in a staged process (Seitz and Lisch 2011), which addresses symptoms, but not the underlying cause. The age on onset and severity of the dystrophic phenotype can vary substantially, and the treatment strategy is tailored to the individual patient (Klintworth 2009). Corneal dystrophies may remain asymptomatic and undiagnosed in the absence of trauma (Chao-Shern et al. 2018; Nithianandan et al. 2019). In other instances, opacities begin to appear in early childhood, leading to reduced visual acuity and ultimately resulting in legal blindness. For milder cases, conservative therapies such as gels/ointments and therapeutic contact lenses are sufficient. More severe cases require surgical interventions that are based on the location of the opacities within the cornea (Seitz and Lisch 2011).

Laser Eye Surgery

Recurrent corneal erosions and corneal dystrophies involving the anterior stroma can be treated with phototherapeutic keratectomy (PTK), which delays the need for corneal transplantation (NIH 2016). PTK can be considered for superficial epithelial dystrophies and, with less success, stromal dystrophies. The technique uses an excimer laser to etch the cornea precisely, with minimal adjacent tissue damage. The ablated surface epithelium regrows rapidly to form normal basal lamina complexes over the cornea (Steinert 2013). This approach, however, addresses only symptoms and not their underlying cause. It is not curative and often offers only a temporary solution, typically requiring repeated treatments (Dinh et al. 1999; Hafner et al. 2005; Chen and Xie 2013; Rathi et al. 2016).

Corneal Transplantation

Unfortunately, in many instances sight deteriorates to the point at which a corneal transplantation is required. While well-established, drawbacks including the shortage of corneal donors in some areas of the world and graft rejection following transplant limit its use (Bidaut-Garnier et al. 2016). This procedure removes the diseased central cornea and replaces it with a healthy donor cornea. It restores vision and reduces pain; however, permanent weakening of the cornea, the risk of rejection, unpredictable refractive errors, cost, and the need for continued follow-up indefinitely make it a less-than-ideal treatment—especially in areas of the world where good medical care is not available.

The benefit of corneal grafts for corneal dystrophies related to *TGFBI* mutations is often short-lived because of recurrence of the underlying corneal dystrophy. Disease recurrence is due to repopulation of the graft by host epithelial and stromal cells (Jun et al. 2004; Aldave et al. 2007; Kim et al. 2008; Han et al. 2016; Jun et al. 2018; Lang et al. 2015). It is not uncommon for patients with corneal dystrophies to undergo multiple corneal transplants during their lifetime. A minimally invasive approach that addresses the underlying genetic cause of corneal dystrophy would present a very attractive treatment option (Christie et al. 2017).

Pharmacological Treatments

New pharmacological treatments that may inhibit or delay amyloid deposition are currently in early stages of development.

Chaperone Therapy

The production of mutant transforming growth factor beta-induced protein (TGFBIp) involves proteolytic processing and clearance processes that are different from those of wild-type protein. This difference could be exploited if compounds that only bind to mutant TGFBIp and its unique binding sites could be identified. Osmolytes are small organic molecules that regulate solvent properties of cells (Diamant et al. 2001). They are widely used as chemical chaperones that can stabilize and facilitate protein folding (Welch et al. 1996). In an in vitro study, four osmolytes were tested (Betaine, Raffinose, Sarcosine, and Taurin). They inhibited the formation of amyloid fibrils and enhanced their disaggregation without producing any adverse effects (Venkatraman et al. 2020). This means that they have a potential to progress to in vivo studies once suitable animal models are developed.

Pharmaceutical Modulation

Venkatraman et al. (2020) were first to screen a large number of compounds for a potential therapeutic effect against mutant TGFBIp. Over 2500 compounds were screened, and binding sites were validated. Three lead compounds were found to be effective in preventing generation of amyloidogenic peptides from the H572R mutant protein. This initial study, which only targeted R555W and H572R mutants, identified potentially effective compounds that might delay or prevent mutant TGFBIp aggregation. They remain in the very early stages of development. In vitro and in vivo models have yet to be created to further this work.

Gene Therapy

The cornea is an ideal candidate for targeted gene therapy due to its small surface area, accessibility, and ease of visualization (Moore et al. 2018). In addition, the unique immune privileged status of the cornea (Hori et al. 2019) minimizes immune response to the gene-based therapeutics and delivery vehicles that might be introduced. Furthermore, the avascularity of the cornea minimizes the risk that any of the administered gene therapy and delivery agents will escape the cornea to reach extraocular sites and cause unintended adverse events in other tissues.

Conventionally, gene therapy refers to the introduction of a functional copy of the gene, usually as a viral transgene, to treat loss of function mutations (Fig. 10.2). This allows the target cell to produce a functional protein when the endogenous protein is defective. Such an approach has been successfully applied to treat recessive, loss-of-function, heritable diseases of the retina (Bennett 2017). For example, Leber's congenital amaurosis type II (LCA2), due to loss-of-function mutations in the *RPE65* gene, has been treated by subretinal injection of *RPE65* cDNA packaged in AAV2 (Bennett 2017) and was the first such reagent to receive FDA approval (Darrow 2019).

Such an approach for the treatment of dominant disorders is more problematic since expression of the mutant gene product persists alongside the expression of the functional transgene (Fig. 10.3). The success of this approach may depend upon the mechanism of action of the dominant mutation.

For instance, milder forms of retinitis pigmentosa (RP) may be caused by dominant-negative mutations in which the mutant protein dimerizes with normal rhodopsin and interferes with its function or assembly in the disc membrane. In contrast, more severe forms of RP may be caused by gain-of-function mutations toxic to the rod cell either through mis-localization or by continuous activation of the unfolded protein response (Lewin et al. 2014). The balance between functional homodimers and nonfunctional dominant-negative rhodopsin heterodimers may be changed by increasing the level of normal protein through transgene expression (Lewin et al. 2014).

The expression of injurious mutant gain-of-function mutations may need to be suppressed to achieve therapeutic benefit from the virally delivered transgene. Suppression of expression may be achieved in an allele-specific manner (LaVail et al. 2000) or by blocking expression of both endogenous wild-type and mutant alleles and delivery of a suppression-resistant replacement transgene (Ader et al. 2005; Millington-Ward et al. 2011).

Meesmann epithelial corneal dystrophy (MECD) has been shown to result from cellular toxicity caused by mutant keratin 12 protein induction of the unfolded protein response (Allen et al. 2016), while mutant TGFBI protein has been shown to accumulate in the corneal deposits seen in granular and lattice *TGFBI* corneal dystrophies (TGFBICD) (Courtney et al. 2015). This suggests that, in both cases, suppression of expression of the mutant protein is required.

The autosomal dominant inheritance pattern of the *TGFBI* corneal dystrophies suggests that selective ablation of only the mutant allele may be a viable treatment strategy (Christie et al. 2017; Moore et al. 2018). Heterozygous nonsense mutations in *TGFBI* can produce a normal phenotype (Sakimoto et al. 2003; Song et al. 2015), indicating that *TGFBI* is haplosufficient and disruption of the mutant allele alone would not lead to a more severe phenotype. Both hetero-

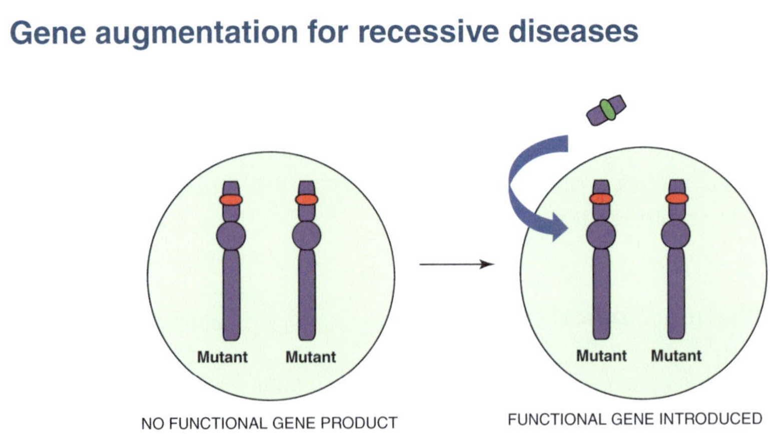

Fig. 10.2 In recessive disorders, introduction of a transgene encoding wild-type protein allows the target cell to produce a functional protein in cases when the endogenous protein is absent or defective

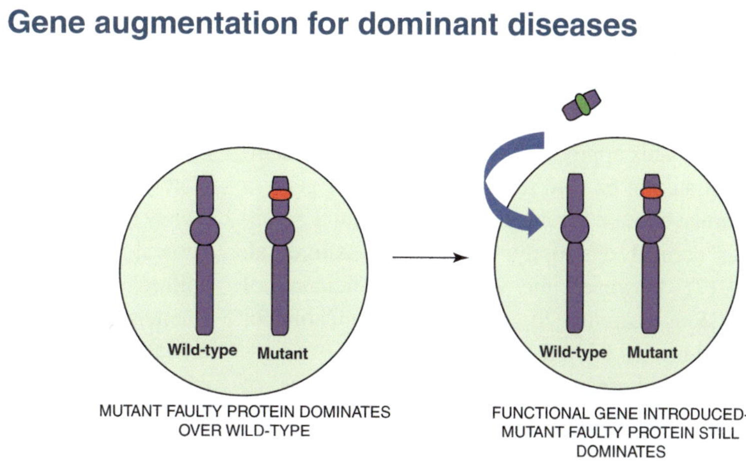

Fig. 10.3 In dominant disorders, expression of the mutant gene product persists alongside the expression of the functional protein by the introduced transgene

zygous null *TGFBI* and *Krt12* mice have been reported to be phenotypically normal (Ahlfeld et al. 2016; Kao et al. 1996). However, while the homozygous null *TGFBI* mouse has been shown to display only small changes in corneal structure (Poulsen et al. 2018), wound healing in a zebrafish tail fin regeneration assay was inhibited following *tgfbi* morpholino injection (unpublished observations) indicating an essential role of TGFBIp in wound repair and suggesting maintenance of wild-type TGFBIp in the cornea is required to maintain normal repair responses. Complete knockout of *TGFBI* could be detrimental should the cornea be injured (when TGFBI expression is known to be upregulated) and thus allele-specific gene disruption is necessary, in which only the mutant allele is targeted for disruption, leaving the wild-type allele intact.

siRNA

The potential of therapeutic small inhibitory RNA (siRNA) therapy to suppress mutant protein expression for autosomal dominant corneal dystrophies has been clearly demonstrated both for MECD and TGFBICD (McLean and Moore 2011; Liao et al. 2011; Allen et al. 2013; Courtney et al. 2014a, b). Alternatively, bi-allelic knock down of *TGFBI* expression has been demonstrated in human corneal cell lines (Yuan et al. 2007; Yellore et al. 2011). However, siRNA-induced knockdown of mutant protein expression would require daily application of siRNA, and uptake of therapeutic levels of these charged molecules by the intact cornea is problematic but possible (Schiroli et al. 2019; Baran-Rachwalska et al. 2020). Furthermore, for a mutant allele-specific approach to be pursued, development and testing of a different siRNA would be required for each of the known >70 *TGFBI*, 23 *KRT12*, and 3 *KRT3* mutations.

CRISPR/Cas9 Gene Editing

CRISPR/Cas9 gene editing provides a gene therapy approach that could provide a permanent or longer lasting suppression of mutant protein expression. Cas9 is a bacterial ribonucleoprotein endonuclease whose DNA target specificity is determined by an RNA molecule known as the guide sequence (Ran et al. 2013). Cas9 can be directed to cut a desired sequence in the genome, provided it is directly upstream of a protospacer adjacent motif (PAM), determined by the particular Cas9 protein used, by simply altering the guide RNA sequence. The site-specific sgRNA will direct the Cas9 nuclease to make a double-strand break (DSB) and the cell will then attempt to repair this damage, by either precise homology directed repair (HDR) or error-prone non-homologous end joining (NHEJ) (Ran et al. 2013). By harnessing these cellular responses, different forms of gene editing can be achieved.

In HDR, the DSB is repaired using an exogenous homologous repair template, added together in large excess with the CRISPR/Cas9 gene editing components, to encourage repair mechanisms to introduce desired sequence changes and achieve gene correction. However, as homologous recombination is restricted to the S and G2 phase of the cell cycle, gene correction via HDR is limited to dividing cells (Iyama and Wilson 2013). This approach has been successfully demonstrated in ex vivo repair of the *TGFBI* gene in human corneal keratocytes derived from a granular corneal dystrophy patient (Taketani et al. 2017). However, the achievable efficiency of HDR is currently too low to be applicable for local in vivo administration.

Two alternative approaches to the use of NHEJ-induced repair errors by CRISPR/Cas9 in the treatment of corneal dystrophies can be envisaged (Fig. 10.4). In the first, bi-allelic gene disruption abolishes expression from both wild-type and mutant alleles. A guide sequence can be chosen that can be used for all known mutations using criteria of high on-target and very low off-target cutting in which the guide sequence does not overlap with any of the known mutations. Such an approach has been used to generate *TGFBI* knockout limbal epithelial stem cells in which bi-allelic disruption of *TGFBI* has been demonstrated (Kim et al. 2019). In practice, a limbal biopsy containing limbal epithelial stem cells (LESCs) would be taken from the patient, the LESCs would be genetically altered ex vivo and reintroduced into the patient's cornea. However, in order for the clonal edited cells to repopulate the cornea, resident stem cells still bearing the mutation would have to be completely removed, a state of limbal stem cell deficiency (LSCD), a severe condition which leads to corneal conjunctivalization, neovascularization, epithelium instability, and visual loss (Fernandez-Buenaga et al. 2018). The risk that the graft of clonally edited LESCs would fail may be considered too high; thus, this would not be a feasible treatment strategy for the corneal dystrophies. However, an approach that selectively disrupts the mutant allele and leaves the wild-type allele intact would be a viable treatment option.

Mutation-dependent allele-specific genome editing can be achieved by harnessing the mechanism of action of the RNA-guided Cas9 nuclease. The Cas9 nuclease can be redirected to a desired target sequence simply by altering the 20-nucleotide spacer sequence within the sgRNA, provided it lies directly upstream of a protospacer adjacent motif (PAM) recognized directly by the Cas9 protein (Jinek et al. 2012; Mojica et al. 2009). PAM recognition is the initial step in Cas9 binding to a target site, the PAM interacting

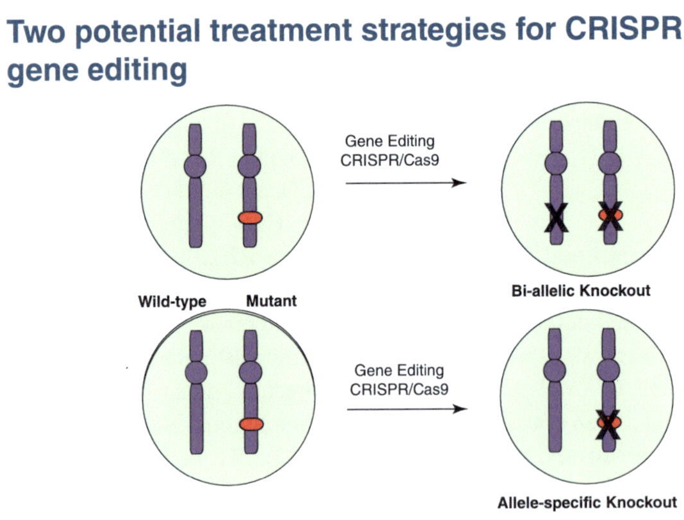

Fig. 10.4 Bi-allelic gene disruption abolishes expression from both the wild-type and mutant alleles. A guide sequence can be chosen that can be used for all known mutations using criteria of high on-target and very low off-target cutting in which the guide sequence does not overlap with any of the known mutations. Mutation-dependent allele-specific genome editing can be achieved by selecting guides that place the mutation either in the PAM site (PAM-specific) or in the 20 bp sgRNA guide sequence (guide-specific)

domain samples the PAM. This dependence upon PAM recognition was exploited to demonstrate, for the first time, allele specificity of gene editing where the disease-causing missense mutation generated a PAM recognition motif not present in the wild-type sequence (Courtney et al. 2016). Thus, a MECD-causing mutation in *Krt12* results from a single base pair T > C transition mutation. This alters the wild-type sequence CTT (Leu) to the mutant CCT (Pro), and on the opposite strand, this alters the sequence from a weak non-canonical PAM (AAG) to a strong Cas9 PAM (AGG). An sgRNA targeting Cas9 to this site was found to be highly effective at reducing the expression of mutant *KRT12* mRNA and protein in vitro and resulted in NHEJ repair-induced in frame-shifting deletions within the mutant *KRT12* allele when a combined Cas9/sgRNA expression construct was injected into the corneal stroma of a humanized MECD mouse model (Courtney et al. 2016) (Fig. 10.5). This approach was further validated in a lattice corneal dystrophy-associated *TGFBI* mutation (c.1580 T > G, L527R) that generates a novel PAM (CTG > CGG). Specificity was confirmed by both a previously described in vitro dual-luciferase assay (Allen et al. 2013; Courtney et al. 2014a) and by an in vitro DNA cleavage assay (Christie et al. 2017; Moore et al. 2018) (Fig. 10.6). Genome targeting can be improved with engineered CRISPR/Cas9 variants which remove some of the constraints of the PAM recognition mechanism. An engineered *Streptococcus pyogenes* Cas9 (SpCas9) allows targeting to be expanded to NGN PAM sequences, and a further optimized variant called SpRY (NRN > NYN PAMs) allows almost near-PAM-less targeting (Walton et al. 2020). However, this new method has been only investigated in vitro and its in vivo utility is yet to be determined.

An alternative to the PAM-specific approach is offered by the next stage in target recognition by the Cas9-sgRNA ribonucleoprotein. Once the correct PAM has been identified, the region of the guide proximal to the PAM will direct target binding (Anders et al. 2015; Cencic et al. 2014; Sternberg et al. 2014). The PAM proximal region, consisting of the first 5–12 bp of the guide sequence (Pattanayak et al. 2013; Sternberg et al. 2014; Wu et al. 2014), has been coined the "seed region." As the PAM proximal region dictates whether Cas9 binds to the target site, mismatches in this region are critical in determining specificity of Cas9. If the PAM proximal region is complementary to the guide sequence, the PAM distal region will then bind. Although complementarity between the PAM distal region and guide sequence is required, reports have shown mismatches here are better tolerated than in the PAM proximal region (Pattanayak et al. 2013; Sternberg et al. 2014; Wu et al. 2014). Thus, should the disease-causing missense mutation be located within this PAM proximal region, it may be possible to distinguish between the wild-type and mutant alleles and disrupt the mutant allele while leaving the wild-type allele intact.

Over 70 missense mutations have been identified in *TGFBI* that cause a variety of corneal dystrophies (Han et al. 2016) (Fig. 10.7). These mutations span the entire *TGFBI* gene but are clustered in hotspots found in exon 4, 11, 12, and 14. An analysis of the known *TGFBI* missense mutations was performed to determine whether they generate a novel *S. pyogenes* Cas9 PAM or if a *S. pyogenes* Cas9 PAM is found within the eight nucleotides, placing the mutation within the seed region (Christie et al. 2017). 19/60 mutations generate a novel S. pyogenes PAM, while 44/60 have a naturally occurring adjacent PAM site that places the mutation within the seed region. When these observations are considered together, 20% of the *TGFBI* missense mutations cannot be targeted by *S. pyogenes* Cas9 (Fig. 10.8).

Despite the wide spectrum of *TGFBI* mutations, most are found in only one or a few cases, and the vast majority of cases are due to the five most prevalent mutations found in either codon 124 (exon 4) (R124C, R124H, R124L) or codon 555 (exon 12) (R555Q and R555W) (Table 10.2).

Remarkably, each of these mutations, differing by only a single amino acid, results in strikingly different protein aggregates with a very strong genotype-phenotype correlation. None of these five mutations generate a novel PAM; however, the missense mutation is present in the seed region of the guide sequence in all cases. To achieve

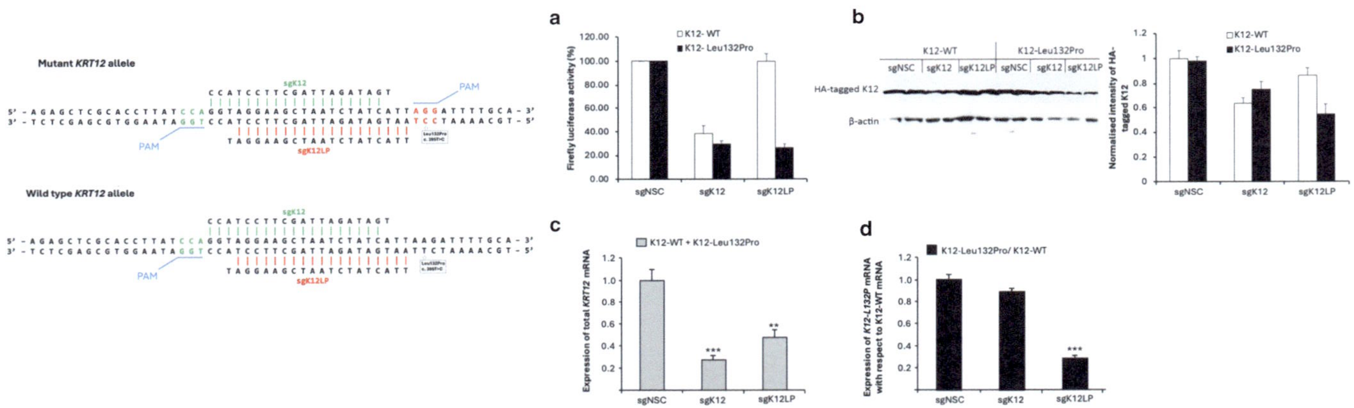

Fig. 10.5 Design of sgRNAs for targeting wild-type and mutant K12 alleles. An sgRNA to use the SNP-derived PAM found on the K12-L132P allele was designed (red). This PAM is absent from the wild-type allele. A second sgRNA to target both wild-type and mutant K12 alleles (green) was also designed and used as a positive control. Evaluation of allele specificity and potency of sgK12LP using exogenous constructs. Exogenous expression constructs for wild-type and mutant K12 were employed to test the allele specificity and potency of sgK12LP. (**a**) A dual luciferase assay demonstrated the allele specificity of the sgK12LP plasmid, whereas potency was shown to be comparable to that of the sgK12 construct. N = 8. (**b**) Western blotting further demonstrated these attributes with a noticeable reduction in K12-L132P protein in cells treated with sgK12LP in comparison with cells treated but expressing K12 wild-type protein. β-Actin was used as a loading control. (**c**) Quantitative reverse transcriptase PCR for total K12 in cells expressing both wild-type and mutant alleles demonstrated a knockdown in mRNA expression. N = 4. (**d**) Allele proportions of this mRNA knockdown were then quantified by pyrosequencing, confirming an allele knockdown of the mutant allele in cells co-expressing both KRT12 alleles and treated with sgK12LP. N = 4, $^{*}P < 0.05$, $^{**}P < 0.01$, $^{***}P < 0.001$

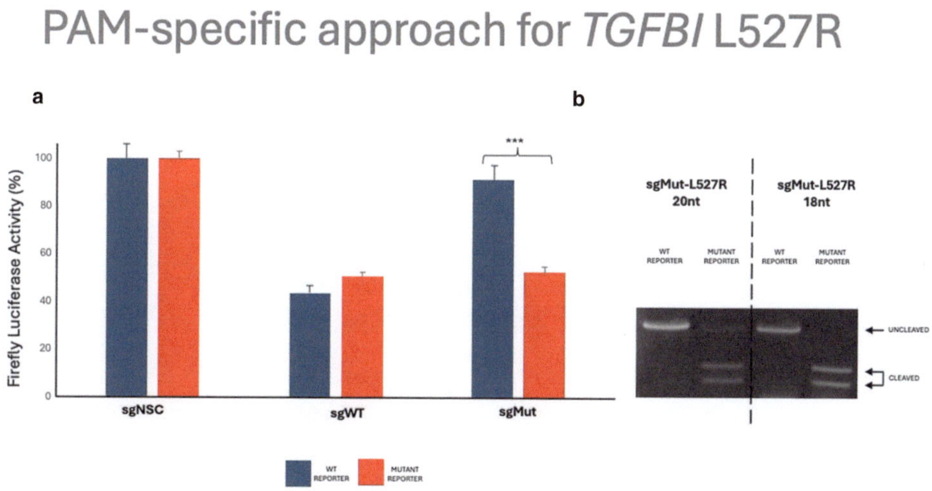

Fig. 10.6 Allele-specific cleavage of L527R *TGFBI* mutation utilizing a PAM-specific approach. (**a**) Both sgWT and sgMUTANT were targeted to a luciferase reporter plasmid containing either a wild-type or mutant *TGFBI* sequence to determine potency and allele specificity. (**b**) An in vitro digestion with Cas9 protein complexed with a sgRNA utilizing the novel L527R PAM was carried out to confirm the specificity observed. Mutant guides of both 20 and 18 nucleotides were tested

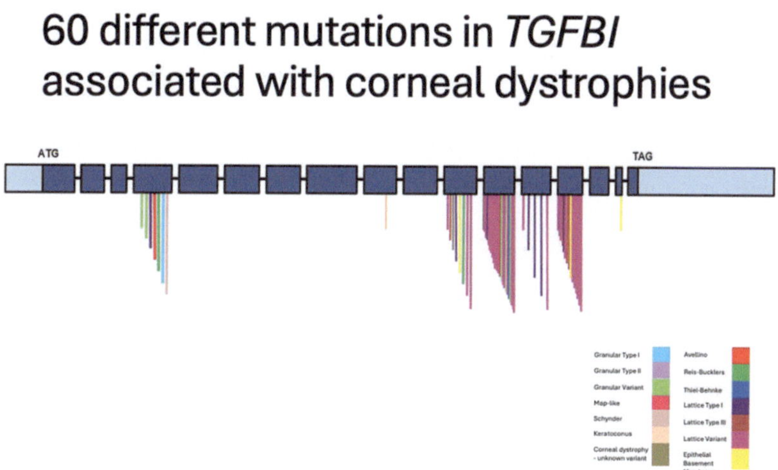

Fig. 10.7 Schematic of the 17 exon *TGFBI* gene. Coding exons are shaded dark blue and noncoding regions are shaded light blue. The location of mutations is indicated below and color coded as shown. Hotspots for mutations are clearly seen in exons 4, 11, 12, and 14

complete allele specificity for a particular *TGFBI* mutation, stringent fidelity is required as an almost perfect off-target site exists in the form of the wild-type allele, which differs by only one base pair from the mutant. Therefore, we investigated if a guide-specific approach could confer stringent allele specificity (Christie et al. 2017). We found that *S. pyogenes* Cas9 was not able to adequately discriminate between wild-type and mutant alleles differing in sequence at only one base of the guide sequence (Fig. 10.9). Reports which indicated that truncating the guide RNA can improve specificity led us to perform a screen of guide length from 16 to 22 bp (Figs. 10.10, 10.11, 10.12, 10.13, and 10.14). However, no significant improvement in specificity was observed. Interestingly, we found that mismatches closer to the PAM conferred more discrimination between

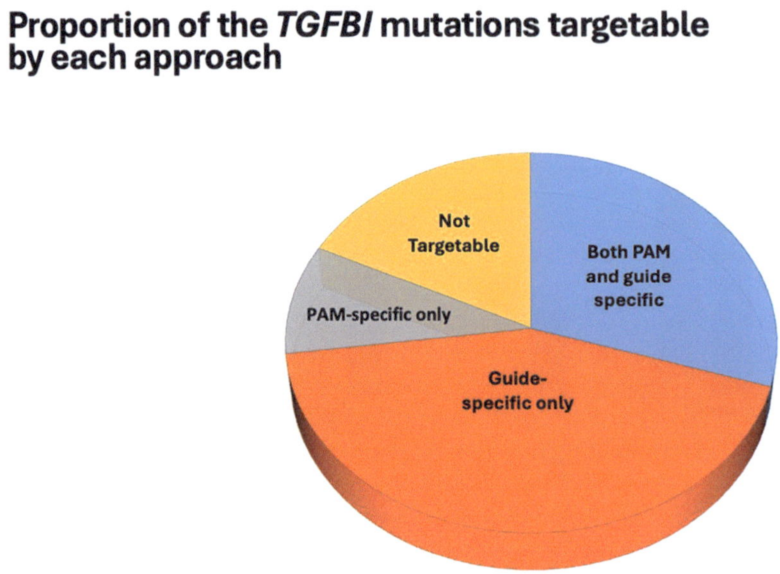

Fig. 10.8 Venn diagram to illustrate the total number of *TGFBI* mutations that (i) generate a novel *S. pyogenes* PAM; (ii) have a nearby *S. pyogenes* PAM, i.e., within the first 8 bp of the guide sequence; (iii) have both a novel and nearby *S. pyogenes* PAM; or (iv) are not targetable by either approach

Table 10.2 Summary of the most common mutations in the *TGFBI* gene and the associated corneal dystrophies

Mutation	Associated corneal dystrophy	Codon change
R124C	Classic lattice type I	CGC > TGC
R124H	Avellino	CGC > CAC
R124L	Reis-Bucklers	CGC > CTC
R555Q	Thiel-Behnke	CGG > CAG
R555W	Classic granular type I	CGG > TGG

alleles that PAM distal mismatches, confirming previous reports (Pattanayak et al. 2013; Sternberg et al. 2014; Wu et al. 2014). However, even though PAM-proximal mismatches result in superior discrimination, they do not confer enough specificity to be suitable for gene therapy.

There are a number of findings that make a mutation-dependent gene editing strategy difficult to pursue. Each of the large number of different *TGFBI* mutations would require a separate, stringently verified guide. Less than one third of *TGFBI* mutations could be targeted by a PAM-specific approach. A guide-specific approach would lack mutant allele specificity. One fifth of *TGFBI* mutations can be targeted by either approach. We have, therefore, investigated the utility of a mutation-independent allele-specific approach. We have already demonstrated that excellent allele specificity can be achieved between wild-type and mutant alleles differing only by a single base pair when a missense mutation generates a novel PAM site. Similarly, single nucleotide polymorphisms (SNPs) can be associated with a PAM on one allele and not on the other (Lessard et al. 2017; Scott and Zhang 2017). This could be used as a proxy for disease-causing mutations if they lie in proximity to the mutation on the same chromosome (*in cis*). Two studies reported the use of the extensive knowledge of the Huntington (*HTT*) gene haplotypes (Lee et al. 2015), encompassing the trinucleotide repeat (TNR) associated with Huntington's disease, to achieve allele-specific disruption of *HTT* (Shin et al. 2016). For *HTT*, screening of only ten SNPs is sufficient to provide haplotype information for 97% of European Huntington's disease patients (Lee et al. 2015). However, in the case of *TGFBI* missense mutations which are spread over a much greater genomic distance, phase cannot be as easily predetermined.

Building on previous work, we developed an approach for the *TGFBI* corneal dystrophies that would enable allele-specific targeting of the *TGFBI* locus of a patient harboring any missense mutation, provided that the patient has a heterozygous haplotype across some or all of the *TGFBI*

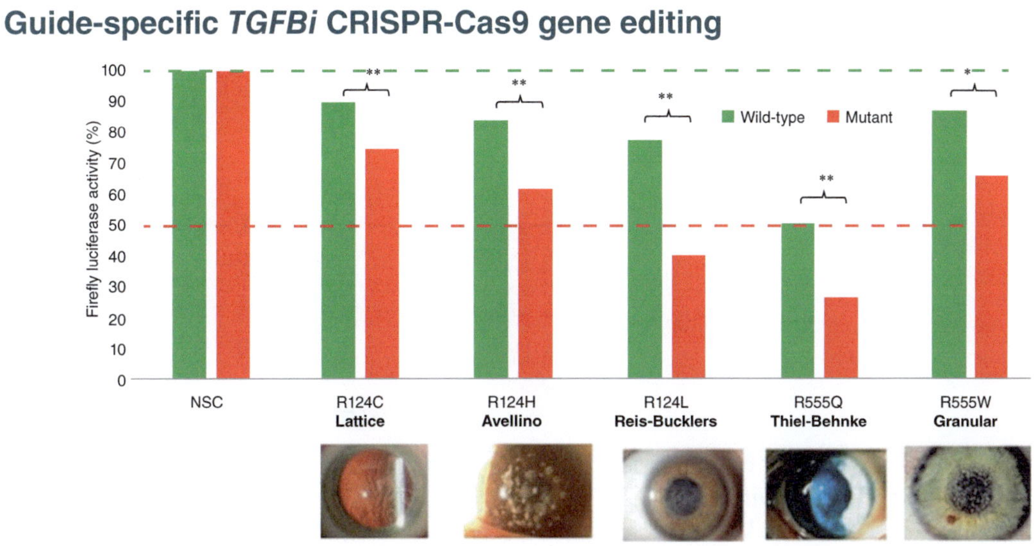

Fig. 10.9 Investigation of a guide-specific approach to treat prevalent *TGFBI* mutations using a guide-specific approach, 20 nucleotide guides for the 5 most prevalent *TGFBI* mutations were targeted to wild-type and respective mutant sequence in a dual luciferase assay. The 5 guides cut with varying degrees of specificities and efficiencies. There was a significant difference between the wild-type and mutant sequence in all cases

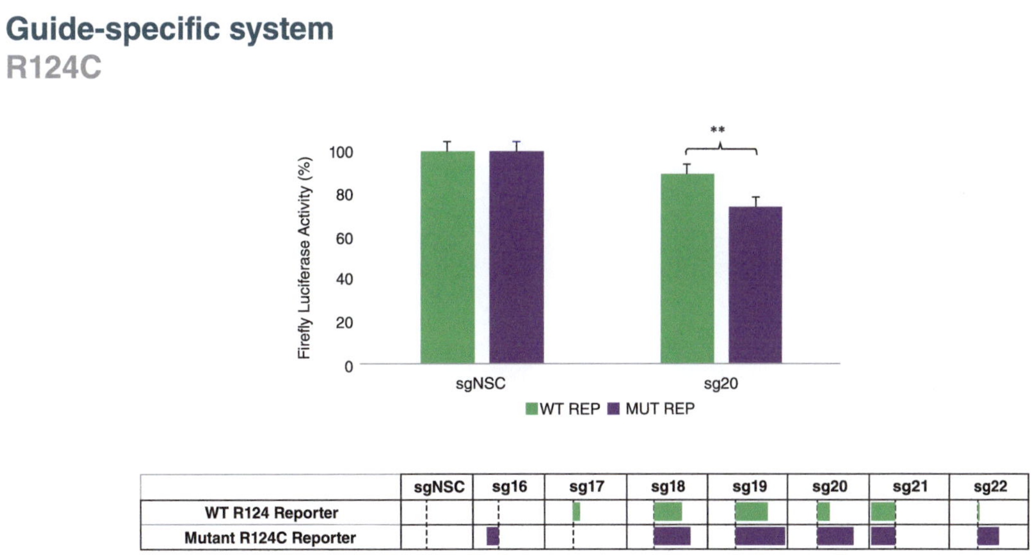

Fig. 10.10 Guide-length screen to determine the effect on specificity of a guide-specific system. Heatmaps showing varying degrees of knockdown observed via a dual luciferase assay when guides ranging in different lengths are targeted to the wild-type and respective mutant sequence. Specificity bars show knockdown when normalized to the nonspecific control, with 100% being maximal knockdown observed

locus. The name coined for this approach was allele-specific, SNP-derived, *in cis*, personalized CRISPR (ASNIP CRISPR) (Fig. 10.15). We identified 24 SNPs across a 35 kbp region containing the *TGFBI* gene that have a minor allele frequency (MAF) of >0.1 in the 1000 Genomes Project Phase 3 cohort (ensuring a reasonable probability that an individual would be heterozygous) and contain a PAM on only one allele. Activity and specificity of guides targeting these SNPs was verified by in vitro DNA cleavage assays. It is possible to determine the phase of

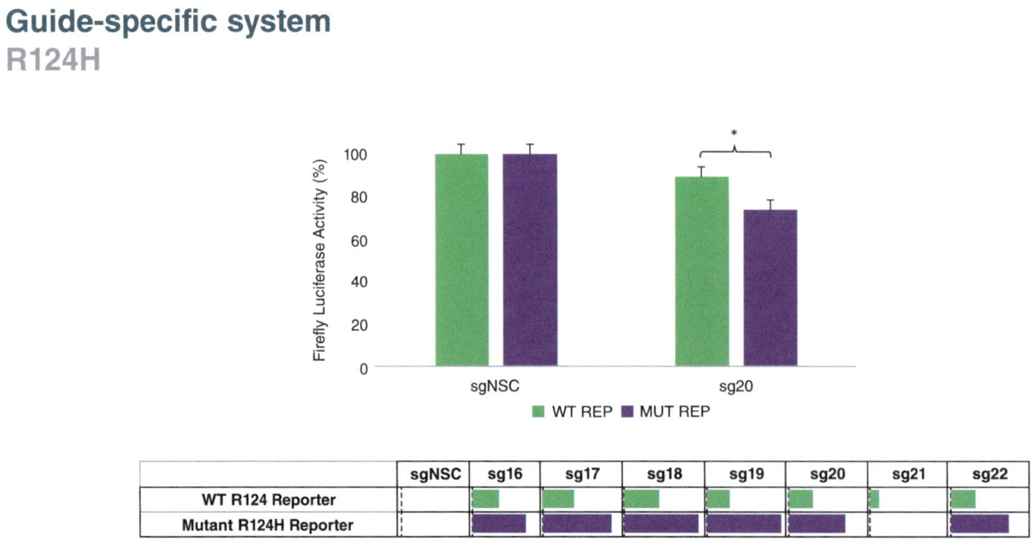

Fig. 10.11 R124H offered an improved specificity in addition to an improved efficiency, although again there was not 100% specificity. The guide length data set again found optimal guide length to range between 18 nt and 20 nt, but no enhanced specificity was observed

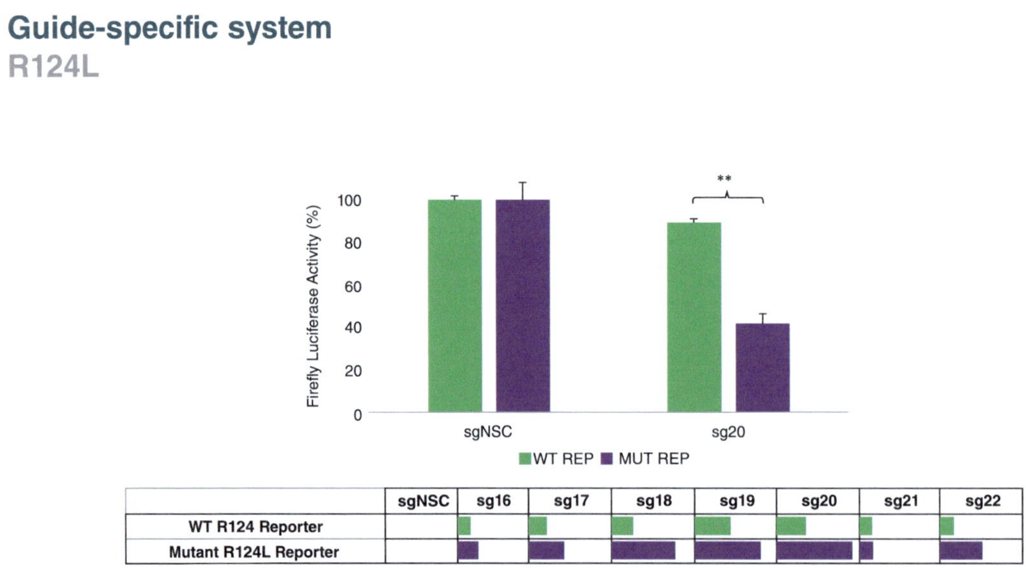

Fig. 10.12 R124L paints the most promising picture with a knockdown of 60% with the 20 nt guide with only a 20% knockdown of the wild-type sequence. Again, although the optimal range appears to be 18 to 20 nt, reducing the guide length has a minimal positive effect on specificity

these PAM-associated SNPs relative to a *TGFBI* missense mutation in a patient by linked-read sequencing (Zheng et al. 2016). Pairwise use of these guides may direct deletion of the intervening region of DNA (Fig. 10.16). Such a dual guide approach to CRISPR/Cas9 editing was shown to be a promising therapeutic approach for the treatment of patients with LCA10 bearing the *CEP290* intronic splice mutation (Ruan et al. 2017), and in a human cell line, it was demonstrated that DNA deletions up to 10 kb coupled to precise NHEJ effectively knocked down a mutant allele (Zheng et al. 2014). Excision of a defined number of exons by this technique gives a much

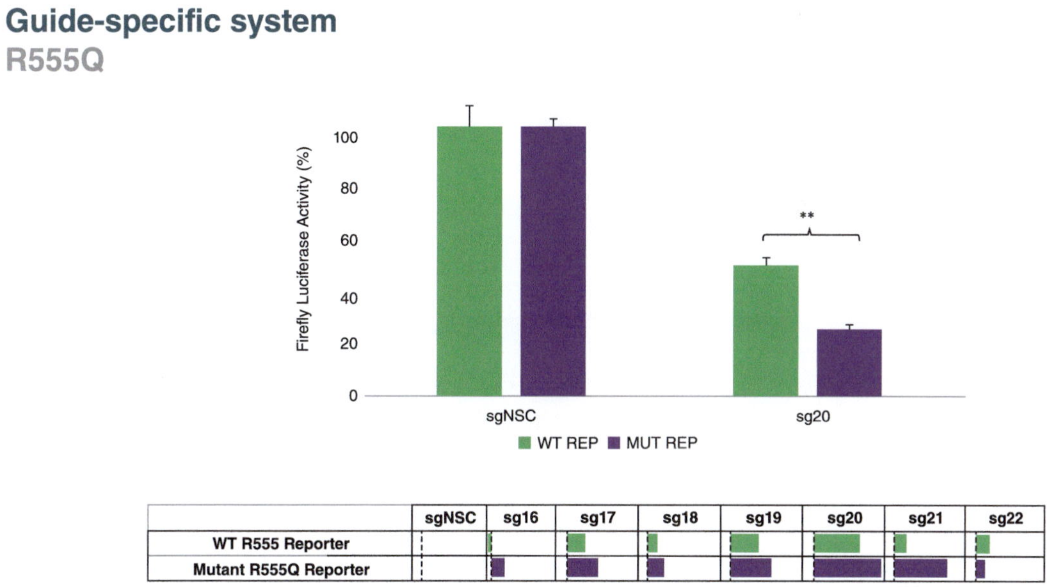

Fig. 10.13 In addition to the R124 mutations, the R555 mutations were investigated in a guide-specific system. Maximal cutting was observed with the 20 nt guide, although the wild-type sequence was dramatically cut, and no allele-specificity can be conferred

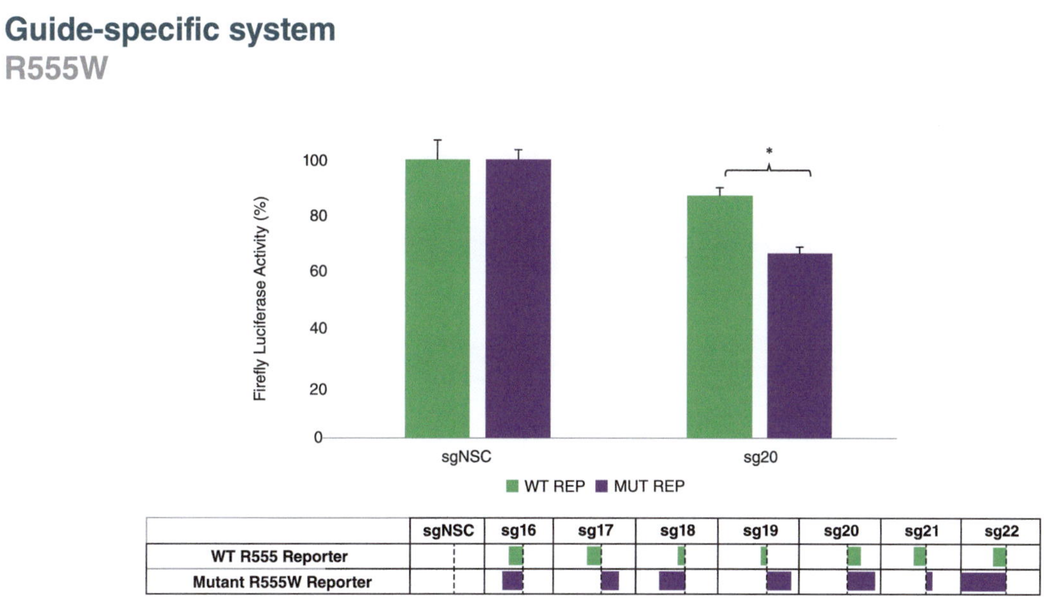

Fig. 10.14 R555W was also tested which did not confer either good specificity or efficiency and in fact offered the least promising outlook

more predictable effect upon the resulting mRNA transcript produced when compared to the unpredictable indel mutations introduced by NHEJ repair of a single double-strand break induced by a single guide. Pairs of guides can be chosen that cause a frameshift mutation, premature termination of translation, and likely nonsense mediated decay of the mutant transcript (Nagy and Maquat 1998; Popp and Maquat 2016).

Deletion of the intervening genomic region was demonstrated by co-transfection of a cell line derived from a patient harboring a R124H *TGFBI*

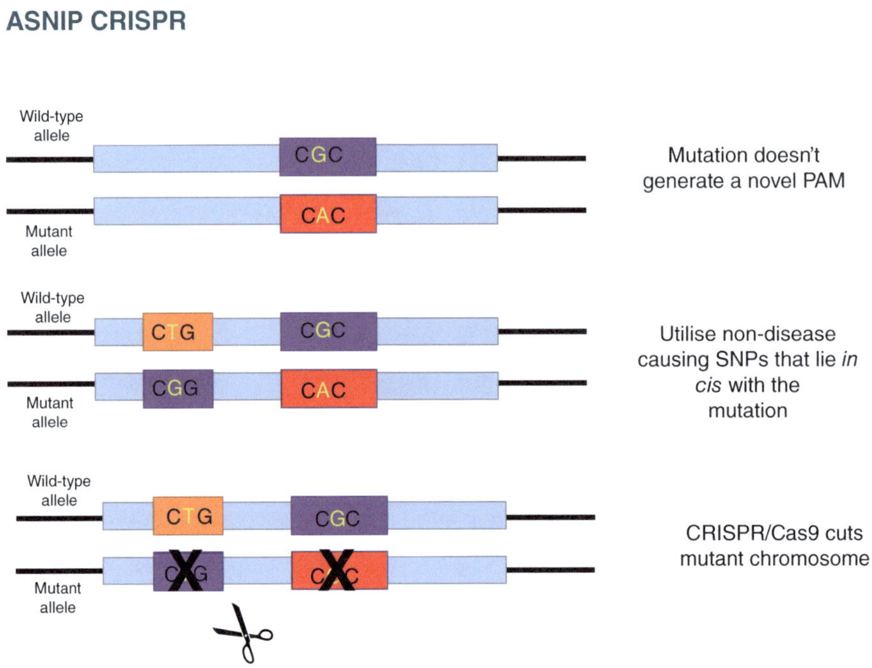

Fig. 10.15 Schematic to illustrate the concept of allele-specific, SNP-derived, in cis, personalized (ASNIP) CRISPR

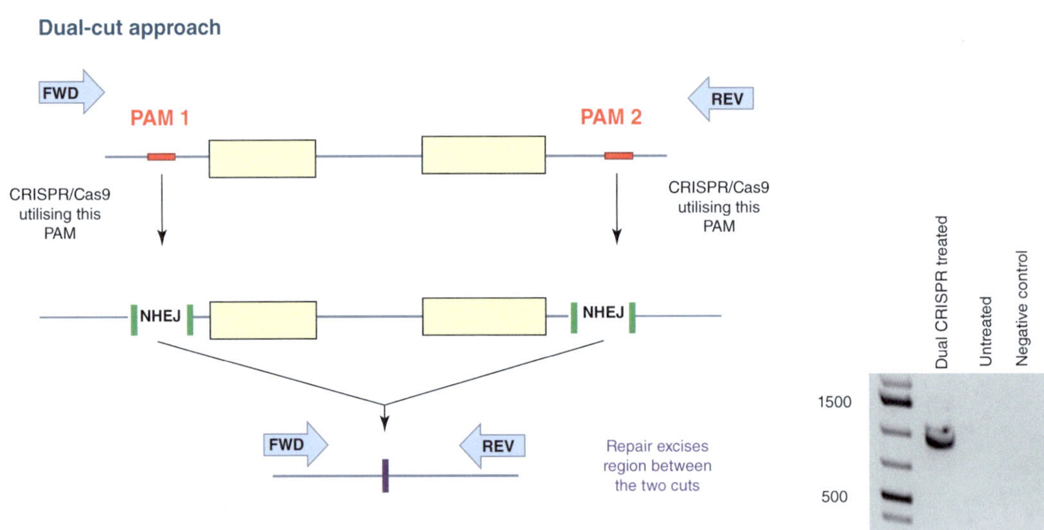

Fig. 10.16 Schematic to show method of detection of deletion of a portion of the TGFBI gene following dual-cut CRISPR/Cas9 gene editing. Prior to, or in the absence of, deletion, Forward and Reverse primers are too far apart for PCR amplification. Deletion of the intervening region, defined by the two PAM sites utilized, brings the primers closer together and this can be detected by PCR

mutation with Cas9-sgRNA RNPs for two of the selected PAM-associated SNPs (Fig. 10.16).

We demonstrated allele-specific CRISPR gene editing, independent of the disease-causing mutation, that is capable of achieving complete allele discrimination. We identified 24 suitable SNPs, but their usefulness would be reduced if PAM-associated alleles all lie together in the same linkage disequilibrium (LD) block. A haplotype analysis was deployed to determine whether all 24 SNPs were suitable for further analysis. The results were further validated by phased sequencing of DNA from a patient harboring a R124H-*TGFBI* mutation where 12 guides were tested and 8 appeared to cleave preferentially the PAM-associated allele, while 4 appeared to have little activity at either the "PAM-associated" or "No PAM present" allele.

It is necessary to demonstrate that selected guides do not cause deleterious alternations elsewhere in the genome. Circularization for in vitro reporting of cleavage effects by sequencing (CIRCLE-seq) was performed to identify any off-targets. Results were further validated with Sanger sequencing and TIDE analysis. For the off-targets that could be analyzed by TIDE, 75% were validated as being true off-targets (18/24). There were 10 off-targets in gene coding regions among the 12 guides tested that were detected by CIRCLE-seq and not by in silico design (Christie et al. 2020). Furthermore, while we have shown encouraging results for the use of a dual cut to induce a therapeutically relevant deletion, Kosicki et al. also reported that complex deletions and rearrangements may also be occurring, which would be undetectable by our PCR-based assays (Kosicki et al. 2018).

Topical Delivery

Despite its accessibility, small size, and immune privileged status, delivery of gene therapy components to the different layers of the cornea has proven to be challenging (Yavuz and Kompella 2017) (Fig. 10.17). The main obstacles to achiev-

Delivering CRISPR/Cas9 components to the cornea

	Plasmid	AAV	RNPs
Topical Application	×	×	×
Intrastromal Injection	✓	✓	✓
Intracameral Injection	×	✓	×

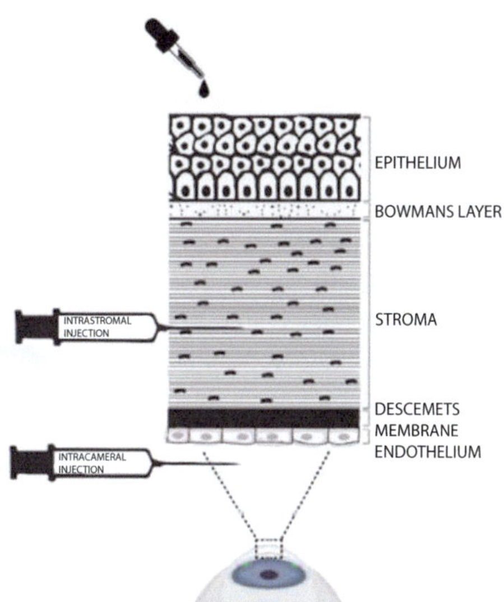

Fig. 10.17 Successful administration routes to the cornea. Schematic of the corneal layers, indicating the location of an intrastromal injection into the stromal layer and an intracameral injection into the anterior chamber

ing a therapeutic dose by diffusion of a drug through the eye surface are the protective mechanisms and underlying ocular anatomy (Guzman-Aranguez et al. 2013; Eghrari et al. 2015; Subrizi et al. 2019). Topically applied drugs may be eliminated from the ocular surface by various mechanisms including lacrimation and tear turnover, drug metabolism, and preferential conjunctival absorption (Kim et al. 2014). The next barrier to entry of gene therapy components is the cornea epithelium in which the cells are linked by characteristically impermeable tight junctions (Hämäläinen et al. 1997). In addition to the biological barriers that exist, the entry of large macromolecules such as DNA or siRNA into corneal cells is challenging. These large, negatively charged, hydrophilic molecules do not easily cross the non-polar cell membrane (Oliveira et al. 2017). Molecules can traverse the corneal epithelium along the transcellular or paracellular pathways, but this requires DNA or siRNA molecules to navigate the exceptionally tight junctions of the corneal epithelium (Hämäläinen et al. 1997). Development of a noninvasive delivery method such as self-administrated eye drops has proven challenging (Li et al. 2014; Souza et al. 2014). Delivery agents need to overcome the tear barrier and remain associated with the corneal epithelium long enough to allow cellular internalization and subsequent release to function within the target cell with optimal bioavailability. Compounds such as NC-1059 can improve the dispersal of riboflavin in the cornea in an ex vivo environment (Zhang et al. 2012).

While commercially available transfection agents such as lipofectamine and polyethyleneimine (PEI) have proven unable to deliver siRNA to mouse cornea in vivo (Li et al. 2014), we have recently demonstrated functional topical delivery of siRNA non-covalently complexed with a modified derivative of a cell penetrating peptide which resulted in significant in vivo knockdown of luciferase reporter gene expression in mouse corneal epithelium (Schiroli et al. 2019). In addition, by using hybrid silicon-lipid nanoparticles (ProSilic), we have demonstrated ProSilic's ability to bind siRNA and deliver functional siRNA to corneal cells both in vivo and in vitro. The siRNA-ProSilic formulation penetrated all corneal layers and resulted in significant knockdown of the Luciferase reporter gene in mice that was sustained for up to 4 days after cessation of application (Fig. 10.18) (Baran-Rachwalska et al. 2020).

However, we have not been able to demonstrate functional delivery of CRISPR-Cas9 gene editing components, either as Cas9-sgRNA RNP, expression plasmid, or viral vector, to the cornea by topical delivery.

Injections

Local administration via intrastromal or intracameral injection (Fig. 10.19) of gene therapy agents circumvents the problem of biological barriers. A reporter knock-in mouse (*Krt12+/luc2*) was used to identify expression levels of *Krt12*. It exclusively expresses firefly luciferase (*luc2*) in the corneal epithelium under control of the keratin K12 promoter. Thus, injection of a co-expression plasmid vector for *Spy*Cas9 and an sgRNA targeting firefly luciferase reporter gene into the corneal stroma significantly reduced luciferase activity from the day after injection, with maximal silencing of 82 ± 13% observed which was also partially sustained for over 7 weeks (Fig. 10.19) (Christie et al. 2017).

Although effective in the small mouse eye, where the pressure achieved during injection can force DNA into cells (Matthaei et al. 2012), when scaled up to a human eye, entry of naked DNA into cells is not particularly effective due to the reduced pressure achievable.

Viral Vectors

The need for pressure transfection can be circumvented using viral vectors to deliver gene editing components. Viral vectors utilized for gene therapy include adenoviruses, adeno-associated viruses (AAV), and retroviruses, specifically lentivirus. Each vector has been extensively tested for use in corneal transduction (Mohan et al. 2005, 2012). No single vector offers all desirable

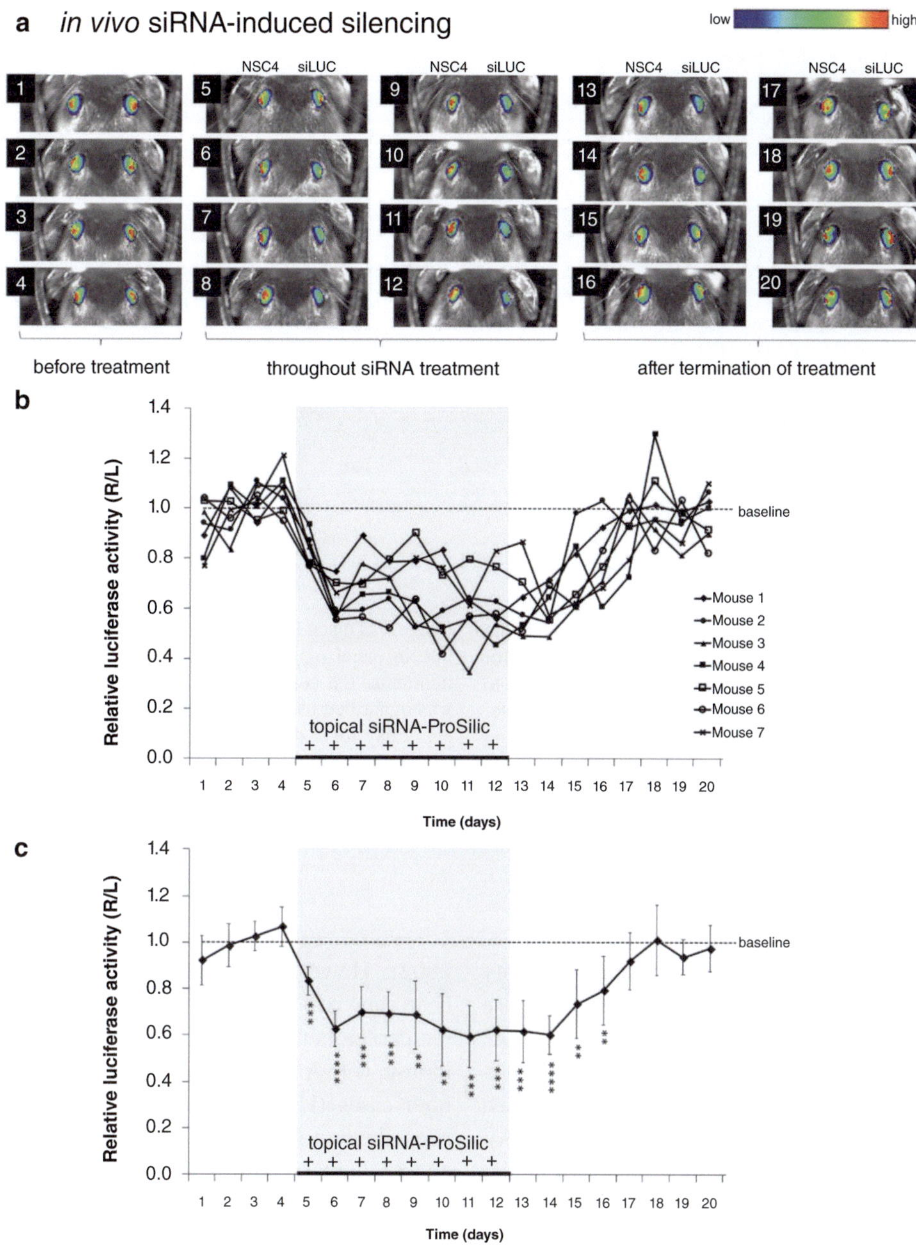

Fig. 10.18 Reporter mice-expressing luciferase specifically in the corneal epithelium was treated with eye drops containing siLUC complexed with ProSilic-DSC613G applied to the right eye (RE). As a negative control non-specific siRNA (NSC4) complexed with the same ProSilic-DSC613G was applied topically to the left eye (LE). The bioluminescent luciferase activity was measured using Xenogen IVIS spectrum system daily throughout treatment regimen, and for a further 8 days after cessation of treatment for monitoring a washout period. Baseline luminescence was defined for each experimental animal by monitoring ocular luciferase activity at 24-h intervals for 4 days prior to treatment. (**a**) Representative daily images of one reporter mice during monitoring the effect of the topical treatment with siRNA complexed with ProSilic-DSC613G. (**b**) Graph showing individual bioluminescence readouts performed on seven reporter mice using split body control, expressed as the right eye/left eye ratio (RE/LE). (**c**) The effect of topical treatment on luciferase reporter gene expression determined for a treatment group ($n = 7$) expressed as an average value of relative RE/LE luciferase bioluminescent activity (siLUC vs. NSC4) ± standard deviation. Significant knockdown of luciferase with siLUC-ProSilic was observed from day 5 to day 16: $^{**}p < 0.01$, $^{***}p < 0.001$

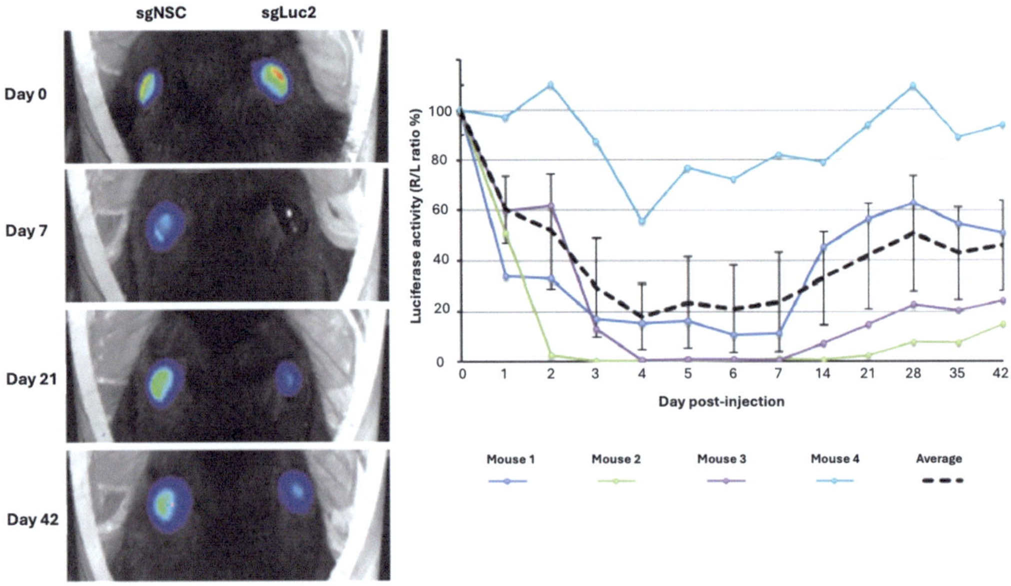

Fig. 10.19 Sustained CRISPR/Cas9-mediated silencing of luc2 in vivo. A short guide RNA (sgRNA) specific for luc2 was designed to target the 5′ region of the gene, to increase the likelihood of inducing a frameshifting deletion that would knock out luciferase activity by generating a premature termination codon. Representative image of mice displaying a maximal reduction in luc2 expression after injection with the sgLuc2 construct (right eye). This image was taken from the mouse represented by the green line in panel (**d**), below, at 7 days posttreatment. After treatment, the corneal luciferase activity of each mouse was quantified using a Xenogen IVIS live animal imager every day for 7 days, then every 7 days thereafter, for a total of 6 weeks. Luciferase activity for each treatment group expressed as a percentage of control (R/L ratio %)

criteria; adenovirus boasts a large cargo capacity but initiates a substantial immune response (Borrás et al. 2001; Bidaut-Garnier et al. 2001; Tsubota et al. 1998), and lentivirus has a considerable packaging capacity but can cause unwanted integration (Athanasopoulos et al. 2017). AAV has been the most widely used vector in ocular gene replacement therapy. AAV provides efficient transduction with a minimal immune response and persistence without integration in the nucleus of non-dividing cells but has a limiting packaging capability (Sharma et al. 2010; Zinn and Vandenberghe 2014). The preferred vector for retinal transduction, AAV-2, does not efficiently transduce the cells of the cornea (Borrás et al. 2002). However, various AAV serotypes have been shown to transduce successfully all layers of the cornea both ex vivo in human and in vivo in mouse, with AAV-2/8 found to be most efficient in both cases (Hippert et al. 2012). AAV-2/9 has been shown to transduce successfully the cornea endothelium in mouse in vivo following intracameral injection (O'Callaghan et al. 2017) and, in our hands, all layers of the cornea by intrastromal injection (Fig. 10.20). Furthermore, systemic injection of AAV-2/9 results in transduction of the corneal epithelium, stroma, and endothelium and appears to transduce limbal epithelial stem cells, as evidenced by sustained centripetal streaks of from the limbus (Basche et al. 2018).

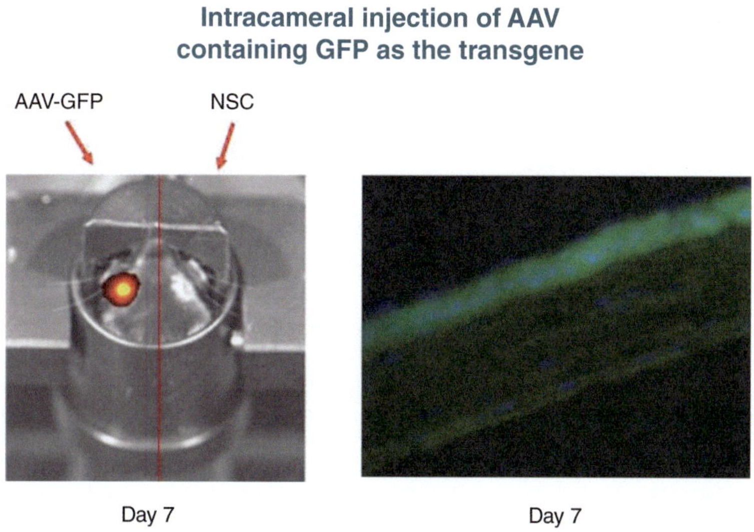

Fig. 10.20 The left eye of a mouse was injected intrastromally with AAV 2/9 containing either a GFP transgene or an empty vector. Seven days after injection, the corneal fluorescence was quantified using a Xenogen IVIS live animal imager. Eyes were collected, fixed in 4% paraformaldehyde in PBS, embedded in paraffin wax, sectioned, and visualized by fluorescence microscopy. GFP fluorescence was observed in all corneal layers from the epithelium (top) to endothelium (bottom)

Conclusion

While considerable progress towards a personalized genome editing strategy to treat the corneal dystrophies has been made, substantial work still remains to be done to overcome the current limitations of CRISPR/Cas9 gene editing, such as a limitation of choice of PAM site, lack of global specificity, and inadequate delivery vehicles. These issues are not isolated to corneal gene therapy but extend to all gene therapy applications. Addressing these outstanding hurdles will enable the safe translation of these CRISPR/Cas9 gene editing therapies to the clinic.

References

Ader M, Kenna PF, Millington-Ward S, Clark G, Kennan A, O'Reilly M, Tam LC, Aherne A, McNally N, Humphries P, Farrar GJ. Toward a gene therapy for dominant disease: validation of an RNA interference-based mutation-independent approach. Mol Ther. 2005;12:555–61.

Ahlfeld SK, Wang J, Gao Y, Snider P, Conway SJ. Initial suppression of transforming growth factor-β signaling and loss of TGFBI causes early alveolar structural defects resulting in bronchopulmonary dysplasia. Am J Pathol. 2016;186(4):777–93.

Afshari NA, Mullally JE, Afshari MA, Steinert RF, Adamis AP, Azar DT, Talamo JH, Dohlman CH, Dryja TP. Survey of patients with granular, lattice, avellino, and Reis-Bücklers corneal dystrophies for mutations in the BIGH3 and gelsolin genes. Arch Ophthalmol. 2001;119(1):16–22. PMID: 11146721.

Aldave AJ, Gutmark JG, Yellore VS, Affeldt JA, Meallet MA, Udar N, Rao NA, Small KW, Klintworth GK. Lattice corneal dystrophy associated with the Ala546Asp and Pro551Gln missense changes in the TGFBI gene. Am J Ophthalmol. 2004;138(5):772–81. https://doi.org/10.1016/j.ajo.2004.06.021. PMID: 15531312.

Aldave AJ, Rayner SA, Kim BT, Prechanond A, Yellore VS. Unilateral lattice corneal dystrophy associated with the novel His572del mutation in the TGFBI gene. Mol Vis. 2006;27(12):142–6. PMID: 16541014.

Aldave AJ, Sonmez B, Forstot SL, Rayner SA, Yellore VS, Glasgow BJ. A clinical and histopathologic examination of accelerated TGFBIp deposition after LASIK in combined granular-lattice corneal dystrophy. Am J Ophthalmol. 2007;143:416–9.

Aldave AJ, Yellore VS, Sonmez B, Bourla N, Salem AK, Khan MA, Rayner SA, Glasgow BJ. A novel variant of combined granular-lattice corneal dystrophy associated with the Met619Lys mutation in the TGFBI gene. Arch Ophthalmol. 2008;126(3):371–7. https://doi.org/10.1001/archopht.126.3.371. PMID: 18332318; PMCID: PMC2860538.

Allen EHA, Atkinson SD, Liao H, Moore JE, Pedrioli DML, Smith FJD, McLean WHI, Moore CBT. Allele-specific siRNA silencing for the common keratin 12 founder mutation in Meesmann epithelial corneal dystrophy. Invest Ophthalmol Vis Sci. 2013;54:494–502.

Allen EH, Courtney DG, Atkinson SD, Moore JE, Mairs L, Poulsen ET, Schiroli D, Maurizi E, Cole C, Hickerson RP, James J, Murgatroyd H, Smith FJ, MacEwen C, Enghild JJ, Nesbit MA, Leslie Pedrioli DM, McLean WH, Moore CB. Keratin 12 missense mutation induces the unfolded protein response and apoptosis in Meesmann epithelial corneal dystrophy. Hum Mol Genet. 2016;25:1176–91.

Anders C, Niewoehner O, Jinek M. In vitro reconstitution and crystallization of Cas9 endonuclease bound to a guide RNA and a DNA target. Methods Enzymol. 2015;558:515–37.

Athanasopoulos T, Munye MM, Yáñez-Muñoz RJ. Nonintegrating gene therapy vectors. Hematol Oncol Clin N Am. 2017;31:753–70.

Atchaneeyasakul LO, Appukuttan B, Pingsuthiwong S, Yenchitsomanus PT, Trinavarat A, Srisawat C. Study Group. A novel H572R mutation in the transforming growth factor-beta-induced gene in a Thai family with lattice corneal dystrophy type I. Jpn J Ophthalmol. 2006;50(5):403–8. https://doi.org/10.1007/s10384-006-0357-6. PMID: 17013691.

Baran-Rachwalska P, Torabi-Pour N, Sutera FM, Ahmed M, Thomas K, Nesbit MA, Welsh M, Moore CBT, Saffie-Siebert SR. Topical siRNA delivery to the cornea and anterior eye by hybrid silicon-lipid nanoparticles. J Control Release. 2020;326:192–202. https://doi.org/10.1016/j.jconrel.2020.07.004. Epub 2020 Jul 9. PMID: 32653503.

Basche M, Kampik D, Kawasaki S, Branch MJ, Robinson M, Larkin DF, Smith AJ, Ali RR. Sustained and widespread gene delivery to the corneal epithelium via in situ transduction of limbal epithelial stem cells, using lentiviral and adeno-associated viral vectors. Hum Gene Ther. 2018;29:1140–52.

Bennett J. Taking stock of retinal gene therapy: looking back and moving forward. Mol Ther. 2017;25:1076–94.

Biber C. Über einige seltenere Hornhauterkrankungen, Inaugural Dissertation. 1890.

Bidaut-Garnier M, Monnet E, Prongué A, Montard R, Gauthier AS, Desmarets M, Mariet AS, Ratajczak C, Binda D, Saleh M, Delbosc B. Evolution of Corneal Graft Survival over a 30-year period and comparison of surgical techniques: a cohort study. Am J Ophthalmol. 2016;163:59–69. https://doi.org/10.1016/j.ajo.2015.12.014. Epub 2015 Dec 17. PMID: 26706619.

Bidaut-Garnier M, Monnet E, Prongué A, Montard R, Gauthier AS, Desmarets M, Borrás T, Gabelt BT, Klintworth GK, Peterson JC, Kaufman PL. Non-invasive observation of repeated adenoviral GFP gene delivery to the anterior segment of the monkey eye in vivo. J Gene Med. 2001;3:437–49.

Borrás T, Gabelt BT, Klintworth GK, Peterson JC, Kaufman PL. Non-invasive observation of repeated adenoviral GFP gene delivery to the anterior segment of the monkey eye in vivo. J Gene Med. 2001;3(5):437–49. https://doi.org/10.1002/jgm.210. PMID: 11601757.

Borrás T, Brandt CR, Nickells R, Ritch R. Gene therapy for glaucoma: treating a multifaceted, chronic disease. Invest Ophthalmol Vis Sci. 2002;43(8):2513–8. PMID: 12147578.

Boutboul S, Black GC, Moore JE, Sinton J, Menasche M, Munier FL, Laroche L, Abitbol M, Schorderet DF. A subset of patients with epithelial basement membrane corneal dystrophy have mutations in TGFBI/BIGH3. Hum Mutat. 2006;27(6):553–7. https://doi.org/10.1002/humu.20331. PMID: 16652336.

Bücklers M. Über eine weitere familiare Hornhautdystrophie. Klin Monatsbl Augenheilkd. 1949;114:386–7.

Cencic R, Miura H, Malina A, Robert F, Ethier S, Schmeing TM, Dostie J, Pelletier J. Protospacer adjacent motif (PAM)-distal sequences engage CRISPR Cas9 DNA target cleavage. PLoS One. 2014;9:e109213.

Chae H, Kim M, Kim Y, Kim J, Kwon A, Choi H, Park J, Jang W, Lee YS, Park SH, Kim MS. Mutational spectrum of Korean patients with corneal dystrophy. Clin Genet. 2016;89(6):678–89. https://doi.org/10.1111/cge.12726. Epub 2016 Feb 10. PMID: 26748743.

Chakravarthi SV, Kannabiran C, Sridhar MS, Vemuganti GK. TGFBI gene mutations causing lattice and granular corneal dystrophies in Indian patients. Invest Ophthalmol Vis Sci. 2005;46(1):121–5. https://doi.org/10.1167/iovs.04-0440. PMID: 15623763.

Chao-Shern C, Me R, DeDionisio LA, Ke BL, Nesbit MA, Marshall J, Moore CBT. Post-LASIK exacerbation of granular corneal dystrophy type 2 in members of a Chinese family. Eye (Lond). 2018;32:39–43.

Chao-Shern C, DeDionisio LA, Jang JH, Chan CC, Thompson V, Christie K, Nesbit MA, McMullen CBT. Evaluation of TGFBI corneal dystrophy and molecular diagnostic testing. Eye (Lond). 2019;33:874–81.

Chen M, Xie L. Features of recurrence after excimer laser phototherapeutic keratectomy for anterior corneal pathologies in North China. Ophthalmology. 2013;120:1179–85.

Cho KJ, Mok JW, Na KS, Rho CR, Byun YS, Hwang HS, Hwang KY, Joo CK. TGFBI gene mutations in a Korean population with corneal dystrophy. Mol Vis. 2012;18:2012–21. Epub 2012 Jul 20. PMID: 22876129; PMCID: PMC3413419.

Christie KA, Robertson LJ, Conway C, Blighe K, DeDionisio LA, Chao-Shern C, Kowalczyk AM, Marshall J, Turnbull D, Nesbit MA, Moore CBT. Mutation-independent allele-specific editing by CRISPR-Cas9, a novel approach to treat autosomal dominant disease. Mol Ther. 2020;28(8):1846–57. https://doi.org/10.1016/j.ymthe.2020.05.002. Epub 2020 May 8. PMID: 32416058; PMCID: PMC7403340.

Christie KA, Courtney DG, DeDionisio LA, Shern CC, De Majumdar S, Mairs LC, Nesbit MA, Moore

CBT. Towards personalised allele-specific CRISPR gene editing to treat autosomal dominant disorders. Sci Rep. 2017;7:16174.

Courtney DG, Atkinson SD, Moore JE, Maurizi E, Serafini C, Pellegrini G, Black GC, Manson FD, Yam GH, Macewen CJ, Allen EH, McLean WH, Moore CB. Development of allele-specific gene-silencing siRNAs for TGFBI Arg124Cys in lattice corneal dystrophy type I. Invest Ophthalmol Vis Sci. 2014a;55:977–85.

Courtney DG, Atkinson SD, Allen EH, Moore JE, Walsh CP, Pedrioli DM, MacEwen CJ, Pellegrini G, Maurizi E, Serafini C, Fantacci M, Liao H, Irvine AD, McLean WH, Moore CB. siRNA silencing of the mutant keratin 12 allele in corneal limbal epithelial cells grown from patients with Meesmann's epithelial corneal dystrophy. Invest Ophthalmol Vis Sci. 2014b;55:3352–60.

Courtney DG, Poulsen ET, Kennedy S, Moore JE, Atkinson SD, Maurizi E, Nesbit MA, Moore CB, Enghild JJ. Protein composition of TGFBI-R124C—and TGFBI-R555W-associated aggregates suggests multiple mechanisms leading to lattice and granular corneal dystrophy. Invest Ophthalmol Vis Sci. 2015;56:4653–61.

Courtney DG, Moore JE, Atkinson SD, Maurizi E, Allen EHA, Pedrioli DML, McLean WHI, Nesbit MA, Moore CBT. CRISPR/Cas9 DNA cleavage at SNP-derived PAM enables both in vitro and in vivo KRT12 mutation-specific targeting. Gene Ther. 2016;23:108–12.

Darrow JJ. Luxturna: FDA documents reveal the value of a costly gene therapy. Drug Discov Today. 2019;24:949–54.

Diamant S, Eliahu N, Rosenthal D, Goloubinoff P. Chemical chaperones regulate molecular chaperones in vitro and in cells under combined salt and heat stresses. J Biol Chem. 2001;276(43):39586–91. https://doi.org/10.1074/jbc.M103081200. Epub 2001 Aug 21. PMID: 11517217.

Dighiero P, Drunat S, Ellies P, D'Hermies F, Savoldelli M, Legeais JM, Renard G, Delpech M, Grateau G, Valleix S. A new mutation (A546T) of the betaig-h3 gene responsible for a French lattice corneal dystrophy type IIIA. Am J Ophthalmol. 2000;129(2):248–51. https://doi.org/10.1016/s0002-9394(99)00324-4. PMID: 10682981.

Dinh R, Rapuan CJ, Cohe EJ, Laibson R. Recurrence of corneal dystrophy after excimer laser phototherapeutic keratectomy. Ophthalmology. 1999;106:1490–7.

Dudakova L, Palos M, Jirsova K, Skalicka P, Dundr P, Liskova P. Novel TGFBI mutation p. (Leu558Arg) in a lattice corneal dystrophy patient. Ophthalmic Genet. 2016;37(4):473–4. https://doi.org/10.3109/13816810.2015.1126615. Epub 2016 Mar 30. PMID: 27028151.

Eghrari AO, Riazuddin SA, Gottsch JD. Overview of the cornea: structure, function, and development. Prog Mol Biol Transl Sci. 2015;134:7–23.

Endo S, Nguyen TH, Fujiki K, Hotta Y, Nakayasu K, Yamaguchi T, Ishida N, Kanai A. Leu518Pro mutation of the beta ig-h3 gene causes lattice corneal dystrophy type I. Am J Ophthalmol. 1999;128(1):104–6. https://doi.org/10.1016/s0002-9394(99)00053-7. PMID: 10482106.

Evans CJ, Davidson AE, Carnt N, Rojas Lopez KE, Veli N, Thaung CM, Tuft SJ, Hardcastle AJ. Genotype-phenotype correlation for TGFBI corneal dystrophies identifies p.(G623D) as a novel cause of epithelial basement membrane dystrophy. Invest Ophthalmol Vis Sci. 2016;57:5407–14.

Fernandez-Buenaga R, Aiello F, Zaher SS, Grixti A, Ahmad S. Twenty years of limbal epithelial therapy: an update on managing limbal stem cell deficiency. BMJ Open Ophthalmol. 2018;3:e000164.

Foja S, Hoffmann K, Auw-Haedrich C, Reinhard T, Rupprecht A, Gruenauer-Kloevekorn C. Identification of two novel mutations in the cornea-specific TGFBI gene causing unique phenotypes in patients with corneal dystrophies. Int Ophthalmol. 2016;36(6):867–73. https://doi.org/10.1007/s10792-016-0216-5. Epub 2016 Mar 10. PMID: 26961680.

Folberg R, et al. Clinically atypical granular corneal dystrophy with pathologic features of lattice-like amyloid deposits. Ophthalmology. 1988;95(1):46–51. https://doi.org/10.1016/S0161-6420(88)33226-4.

Fujiki K, Hotta Y, Nakayasu K, Yokoyama T, Takano T, Yamaguchi T, Kanai A. A new L527R mutation of the betaIGH3 gene in patients with lattice corneal dystrophy with deep stromal opacities. Hum Genet. 1998;103:286–9.

Gruenauer-Kloevekorn C, Clausen I, Weidle E, Wolter-Roessler M, Tost F, Völcker HE, Schulze DP, Heinritz W, Reinhard T, Froster U, Duncker G, Schorderet D, Auw-Haedrich C. TGFBI (BIGH3) gene mutations in German families: two novel mutations associated with unique clinical and histopathological findings. Br J Ophthalmol. 2009;93(7):932–7. https://doi.org/10.1136/bjo.2008.142927. Epub 2008 Nov 10. PMID: 19001012.

Guzman-Aranguez A, Loma P, Pintor J. Small-interfering RNAs (siRNAs) as a promising tool for ocular therapy. Br J Pharmacol. 2013;170:730–47.

Ha NT, Cung le X, Chau HM, Thanh TK, Fujiki K, Murakami A, Kanai A. A novel mutation of the TGFBI gene found in a Vietnamese family with atypical granular corneal dystrophy. Jpn J Ophthalmol. 2003;47(3):246–8. https://doi.org/10.1016/s0021-5155(03)00019-4. PMID: 12782158

Hafner A, Langenbucher A, Seitz B. Long-term results of phototherapeutic keratectomy with 193-nm excimer laser for macular corneal dystrophy. Am J Ophthalmol. 2005;140:1–6.

Hämäläinen KM, Kananen K, Auriola S, Kontturi K, Urtti A. Characterization of paracellular and aqueous penetration routes in cornea, conjunctiva, and sclera. Invest Ophthalmol Vis Sci. 1997;38:627–34.

Han KE, Choi SI, Kim TI, Maeng YS, Stulting RD, Ji YW, Kim EK. Pathogenesis and treatments of TGFBI corneal dystrophies. Prog Retin Eye Res. 2016;50:67–88.

Hippert C, Ibanes S, Serratrice N, Court F, Malecaze F, Kremer EJ, Kalatzis V. Corneal transduction by intra-stromal injection of AAV vectors in vivo in the mouse and ex vivo in human explants. PLoS One. 2012;7:e35318.

Hori J, Yamaguchi T, Keino H, Hamrah P, Maruyama K. Immune privilege in corneal transplantation. Prog Retin Eye Res. 2019;72:100758. https://doi.org/10.1016/j.preteyeres.2019.04.002.

Iyama T, Wilson DM. DNA repair mechanisms in dividing and non-dividing cells. DNA Repair (Amst). 2013;12:620–36.

Jinek M, Chylinski K, Fonfara I, Hauer M, Doudna JA, Charpentier E. A programmable dual-RNA—guided DNA endonuclease in adaptive bacterial immunity. Science. 2012;337:816–22.

Jun RM, Tchah H, Kim TI, Stulting RD, Jung SE, Seo KY, Lee DH, Kim EK. Avellino corneal dystrophy after LASIK. Ophthalmology. 2004;111:463–8.

Jun I, Jung JW, Choi YJ, Kim TI, Seo KY, Kim EK. Long-term clinical outcomes of phototherapeutic keratectomy in corneas with granular corneal dystrophy type 2 excacerbated after LASIK. J Refract Surg. 2018;2018(34):132–9.

Jung JW, Kim SA, Kang EM, Kim TI, Cho HS, Kim EK. Lattice corneal dystrophy type IIIA with hyaline component from a novel A620P mutation and distinct surgical treatments. Cornea. 2014;33(12):1324–31. https://doi.org/10.1097/ICO.0000000000000281. PMID: 25321938.

Kannabiran C, Klintworth GK. TGFBI gene mutations in corneal dystrophies. Hum Mutat. 2006;27(7):615–25. https://doi.org/10.1002/humu.20334. PMID: 16683255.

Kao WW, Liu CY, Converse RL, Shiraishi A, Kao CW, Ishizaki M, Doetschman T, Duffy J. Keratin 12-deficient mice have fragile corneal epithelia. Invest Ophthalmol Vis Sci. 1996;37:2572–84.

Kawasaki S, Yagi H, Yamasaki K, Matsuda A, Takeda K, Kinoshita S. A novel mutation of the TGFBI gene causing a lattice corneal dystrophy with deep stromal involvement. Br J Ophthalmol. 2011;95(1):150–1. https://doi.org/10.1136/bjo.2010.182667. Epub 2010 Oct 24. PMID: 20974628.

Kim T-I, Roh MI, Grossniklaus HE, Kang SJ, Hamilton SM, Schorderet DF, Lee WB, Kim EK. Deposits of transforming growth factor-beta-induced protein in granular corneal dystrophy type II after LASIK. Cornea. 2008;27:28–32.

Kim YC, Chiang B, Wu X, Prausnitz MR. Ocular delivery of macromolecules. J Control Release. 2014;190:172–81.

Kim EK, Kim S, Maeng YS. Generation of TGFBI knockout ABCG2+/ABCB5+ double-positive limbal epithelial stem cells by CRISPR/Cas9-mediated genome editing. PLoS One. 2019;14:e0211864.

Klintworth GK. Lattice corneal dystrophy. An inherited variety of amyloidosis restricted to the cornea. Am J Pathol. 1967;50(3):371–99.

Klintworth GK. Advances in the molecular genetics of corneal dystrophies. Am J Ophthalmol. 1999;128(6):747–54. https://doi.org/10.1016/S0002-9394(99)00358-X.

Klintworth GK. Corneal dystrophies. Orphanet J Rare Dis. 2009;4:7.

Korvatska E, Munier FL, Djemai A, Wang MX, Frueh B, Chiou AG, Uffer S, Ballestrazzi E, Braunstein RE, Forster RK, Culbertson WW, Boman H, Zografos L, Schorderet DF. Mutation hot spots in 5q31-linked corneal dystrophies. Am J Hum Genet. 1998;62:320–4.

Korvatska E, Henry H, Mashima Y, Yamada M, Bachmann C, Munier FL, Schorderet DF. Amyloid and non-amyloid forms of 5q31-linked corneal dystrophy resulting from kerato-epithelin mutations at Arg-124 are associated with abnormal turnover of the protein. J Biol Chem. 2000;275:11465–9.

Kosicki M, Tomberg K, Bradley A. Repair of double-strand breaks induced by CRISPR-Cas9 leads to large deletions and complex rearrangements. Nat Biotechnol. 2018;36:765–71.

Küchle M, et al. Reevaluation of Corneal Dystrophies of Bowman's Layer and the Anterior Stroma (Reis-Bücklers and Thiel-Behnke Types): a light and electron microscopic study of eight corneas and a review of the literature. Cornea. 1995;14(4):333–54. https://doi.org/10.1097/00003226-199507000-00001.

Lakshminarayanan R, Vithana EN, Chai SM, Chaurasia SS, Saraswathi P, Venkatraman A, Rojare C, Venkataraman D, Tan D, Aung T, Beuerman RW, Mehta JS. A novel mutation in transforming growth factor-beta induced protein (TGFβIp) reveals secondary structure perturbation in lattice corneal dystrophy. Br J Ophthalmol. 2011;95(10):1457–62. https://doi.org/10.1136/bjophthalmol-2011-300651. Epub 2011 Aug 10. PMID: 21835759.

Lang SJ, Eberwein P, Reinshagen H, Reinhard T, Sundmacher R. Simultaneous transplantation of limbal stem cells may reduce recurrences of granular dystrophy after corneal transplantation: 2 long-term case reports. Medicine (Baltimore). 2015;94:e789.

LaVail MM, Yasumura D, Matthes MT, Drenser KA, Flannery JG, Lewin AS, Hauswirth WW. Ribozyme rescue of photoreceptor cells in P23H transgenic rats: long term survival and late-stage therapy. Proc Natl Acad Sci. 2000;97:11488–93.

Livshits L, Pampukha VM, Tereshchenko FA, Drozhyna GI. Gene symbol: TGFBI. Disease: corneal dystrophy, lattice type. Hum Genet. 2008;124(3):296–7. PMID: 18846615.

Lee JH, Cristol SM, Kim WC, Chung ES, Tchah H, Kim MS, Nam CM, Cho HS, Kim EK. Prevalence of granular corneal dystrophy type 2 (Avellino corneal dystrophy) in the Korean population. Ophthalmic Epidemiol. 2010;17:160–5.

Lee JM, Kim KH, Shin A, Chao MJ, Abu Elneel K, Gillis T, Mysore JS, Kaye JA, Zahed H, Kratter IH, Daub AC, Finkbeiner S, Li H, Roach JC, Goodman N, Hood L, Myers RH, MacDonald ME, Gusella JF. Sequence-level analysis of the major European Huntington disease haplotype. Am J Hum Genet. 2015;97:435–44.

Lessard S, Francioli L, Alfoldi J, Tardif JC, Ellinor PT, MacArthur DG, Lettre G, Orkin SH, Canver MC. Human genetic variation alters CRISPR-Cas9 on—and off-targeting specificity at therapeutically implicated loci. Proc Natl Acad Sci U S A. 2017;114:E11257–66.

Lewin AS, Rossmiller B, Mao H. Gene augmentation for adRP mutations in RHO. Cold Spring Harb Perspect Med. 2014;4(9):a017400. https://doi.org/10.1101/cshperspect.a017400.

Li Z, Duan F, Lin L, Huang Q, Wu K. A new approach of delivering siRNA to the cornea and its application for inhibiting herpes simplex keratitis. Curr Mol Med. 2014;14:1215–25.

Liao H, Irvine AD, Macewen CJ, Weed KH, Porter L, Corden LD, Gibson AB, Moore JE, Smith FJ, McLean WH, Moore CB. Development of allele-specific therapeutic siRNA in Meesmann epithelial corneal dystrophy. PLoS One. 2011;6:e28582.

Matthaei M, Meng H, Bhutto I, Xu Q, Boelke E, Hanes J, Jun AS. Systematic assessment of microneedle injection into the mouse cornea. Eur J Med Res. 2012;17:19.

Mashima Y, Yamamoto S, Inoue Y, Yamada M, Konishi M, Watanabe H, Maeda N, Shimomura Y, Kinoshita S. Association of autosomal dominantly inherited corneal dystrophies with BIGH3 gene mutations in Japan. Am J Ophthalmol. 2000;130(4):516–7. https://doi.org/10.1016/s0002-9394(00)00571-7. PMID: 11024425.

McLean WH, Moore CB. Keratin disorders: from gene to therapy. Hum Mol Genet. 2011;20(R2):R189–97.

Millington-Ward S, Chadderton N, O'Reilly M, Palfi A, Goldmann T, Kilty C, Humphries M, Wolfrum U, Bennett J, Humphries P, Kenna PF, Farrar GJ. Suppression and replacement gene therapy for autosomal dominant disease in a murine model of dominant retinitis pigmentosa. Mol Ther. 2011;19:642–9.

Mohan RR, Sharma A, Netto MV, Sinha S, Wilson SE. Gene therapy in the cornea. Prog Retin Eye Res. 2005;24:537–59.

Mohan RR, Tovey JCK, Sharma A, Tandon A. Gene therapy in the cornea: 2005-present. Prog Retin Eye Res. 2012;31:43–64.

Mojica FJM, Díez-Villaseñor C, García-Martínez J, Almendros C. Short motif sequences determine the targets of the prokaryotic CRISPR defence system. Microbiology. 2009;155:733–40.

Moore CBT, Christie KA, Marshall J, Nesbit MA. Personalised genome editing—the future for corneal dystrophies. Prog Retin Eye Res. 2018;65:147–65.

Munier FL, Korvatska E, Djemaï A, Le Paslier D, Zografos L, Pescia G, Schorderet DF. Kerato-epithelin mutations in four 5q31-linked corneal dystrophies. Nat Genet. 1997;15(3):247–51. https://doi.org/10.1038/ng0397-247. PMID: 9054935.

Munier FL, Frueh BE, Othenin-Girard P, Uffer S, Cousin P, Wang MX, Héon E, Black GC, Blasi MA, Balestrazzi E, Lorenz B, Escoto R, Barraquer R, Hoeltzenbein M, Gloor B, Fossarello M, Singh AD, Arsenijevic Y, Zografos L, Schorderet DF. BIGH3 mutation spectrum in corneal dystrophies. Invest Ophthalmol Vis Sci. 2002;43(4):949–54. PMID: 11923233.

Musch DC, Niziol LM, Stein JD, Kamyar RM, Sugar A. Prevalence of corneal dystrophies in the United States: estimates from claims data. Invest Ophthalmol Vis Sci. 2011;52:6959–63.

Nagy E, Maquat LE. A rule for termination-codon position within intron-containing genes: when nonsense affects RNA abundance. Trends Biochem Sci. 1998;23:198–9.

NIH. Facts about the cornea and corneal disease. NIH website. 2016.

Niel-Butschi F, Kantelip B, Iwaszkiewicz J, Zoete V, Boimard M, Delpech M, Bourges JL, Renard G, D'Hermies F, Pisella PJ, Hamel C, Delbosc B, Valleix S. Genotype-phenotype correlations of TGFBI p.Leu509Pro, p.Leu509Arg, p.Val613Gly, and the allelic association of p.Met502Val-p.Arg555Gln mutations. Mol Vis. 2011;17:1192–202. Epub 2011 May 5. PMID: 21617751; PMCID: PMC3102024.

Nithianandan H, Chao-Shern C, DeDionisio L, Moore T, Chan CC. Trauma-induced exacerbation of epithelial-stromal TGFBI lattice corneal dystrophy. Can J Ophthalmol. 2019;54:e47–9.

O'Callaghan J, Crosbie DE, Cassidy PS, Sherwood JM, Flügel-Koch C, Lütjen-Drecoll E, Humphries MM, Reina-Torres E, Wallace D, Kiang AS, Campbell M, Stamer WD, Overby DR, O'Brien C, Tam LCS, Humphries P. Therapeutic potential of AAV-mediated MMP-3 secretion from corneal endothelium in treating glaucoma. Hum Mol Genet. 2017;26:1230–46.

Okada M, Yamamoto S, Tsujikawa M, Watanabe H, Inoue Y, Maeda N, Shimomura Y, Nishida K, Quantock AJ, Kinoshita S, Tano Y. Two distinct kerato-epithelin mutations in Reis-Bücklers corneal dystrophy. Am J Ophthalmol. 1998;126(4):535–42. https://doi.org/10.1016/s0002-9394(98)00135-4. PMID: 9780098.

Ołdak M, Szaflik JP, Ścieżyńska A, Udziela M, Maksym RB, Rymgayłło-Jankowska B, Hofmann-Rummelt C, Menzel-Severing J, Płoski R, Żarnowski T, Kruse FE, Szaflik J. Late-onset lattice corneal dystrophy without typical lattice lines caused by a novel mutation in the TGFBI gene. Cornea. 2014;33(3):294–9. https://doi.org/10.1097/ICO.0000000000000062. PMID: 24473223.

Oliveira AV, Rosa da Costa AM, Silva GA. Non-viral strategies for ocular gene delivery. Mater Sci Eng C Mater Biol Appl. 2017;77:1275–89.

Paliwal P, Sharma A, Tandon R, Sharma N, Titiyal JS, Sen S, Kaur P, Dube D, Vajpayee RB. TGFBI mutation screening and genotype-phenotype correlation in north Indian patients with corneal dystrophies. Mol Vis. 2010;29(16):1429–38. PMID: 20680100; PMCID: PMC2913140.

Pang CP, Lam DS. Differential occurrence of mutations causative of eye diseases in the Chinese population. Hum Mutat. 2002;19(3):189–208. https://doi.org/10.1002/humu.10053. PMID: 11857735.

Pattanayak V, Lin S, Guilinger JP, Ma E, Doudna JA, Liu DR. High-throughput profiling of off-target DNA cleavage reveals RNA-programmed Cas9 nuclease specificity. Nat Biotechnol. 2013;31:839–43.

Popp MW, Maquat LE. Leveraging rules of nonsense-mediated mRNA decay for genome engineering and personalized medicine. Cell. 2016;165:1319–22.

Poulsen ET, Runager K, Nielsen NS, Lukassen MV, Thomsen K, Snider P, Simmons O, Vorum H, Conway SJ, Enghild JJ. Proteomic profiling of TGFBI-null mouse corneas reveals only minor changes in matrix composition supportive of TGFBI knockdown as therapy against TGFBI-linked corneal dystrophies. FEBS J. 2018;285:101–14.

Ran FA, Hsu PD, Wright J, Agarwala V, Scott DA, Zhang F. Genome engineering using the CRISPR-Cas9 system. Nat Protoc. 2013;8:2281–308.

Rathi VM, Taneja M, Murthy SI, Bagga B, Vaddavalli PK, Sangwan VS. Phototherapeutic keratectomy for recurrent granular dystrophy in postpenetrating. Indian J Ophthalmol. 2016;64(2):140–4.

Rozzo C, Fossarello M, Galleri G, Sole G, Serru A, Orzalesi N, Serra A, Pirastu M. A common beta ig-h3 gene mutation (delta f540) in a large cohort of Sardinian Reis Bücklers corneal dystrophy patients. Mutations in brief no. 180. Online. Hum Mutat. 1998;12(3):215–6. PMID: 10660331.

Ruan GX, Barry E, Yu D, Lukason M, Cheng SH, Scaria A. CRISPR/Cas9-mediated genome editing as a therapeutic approach for leber congenital amaurosis 10. Mol Ther. 2017;25:331–41.

Sakimoto T, Kanno H, Shoji J, Kashima Y, Nakagawa S, Miwa S, Sawa M. A novel nonsense mutation with a compound heterozygous mutation in TGFBI gene in lattice corneal dystrophy type I. Jpn J Ophthalmol. 2003;47:13–7.

Schiroli D, Gómara MJ, Maurizi E, Atkinson SD, Mairs L, Christie KA, Cobice DF, McCrudden CM, Nesbit MA, Haro I, Moore T. Effective in vivo topical delivery of siRNA and gene silencing in intact corneal epithelium using a modified cell penetrating peptide. Mol Ther Nuc Acids. 2019;17:891–906.

Schmitt-Bernard CF, Guittard C, Arnaud B, Demaille J, Argiles A, Claustres M, Tuffery-Giraud S. BIGH3 exon 14 mutations lead to intermediate type I/IIIA of lattice corneal dystrophies. Invest Ophthalmol Vis Sci. 2000;41(6):1302–8. PMID: 10798644.

Scott DA, Zhang F. Implications of human genetic variation in CRISPR-based therapeutic genome editing. Nat Med. 2017;23:1095–101.

Seitz B, Lisch W. Stage-related therapy of corneal dystrophies. Dev Ophthalmol. 2011;48:116–53.

Sharma A, Tovey JCK, Ghosh A, Mohan RR. AAV serotype influences gene transfer in corneal stroma in vivo. Exp Eye Res. 2010;91:440–8.

Shin JW, Kim KH, Chao MJ, Atwal RS, Gillis T, MacDonald ME, Gusella JF, Lee JM. Permanent inactivation of Huntington's disease mutation by personalized allele-specific CRISPR/Cas9. Hum Mol Genet. 2016;25(20):4566–76. https://doi.org/10.1093/hmg/ddw286. PMID: 28172889; PMCID: PMC6078600.

Skonier J, et al. cDNA cloning and sequence analysis of βig-h3, a novel gene induced in a human adenocarcinoma cell line after treatment with transforming growth factor-β. DNA Cell Biol. 1992;11(7):511–22. https://doi.org/10.1089/dna.1992.11.511.

Song JS, Lim DH, Chung ES, Chung TY, Ki CS. Mutation analysis of the TGFBI gene in consecutive Korean patients with corneal dystrophies. Ann Lab Med. 2015;35:336–40.

Song Y, Sun M, Wang N, Zhou X, Zhao J, Wang Q, Chen S, Deng Y, Qiu L, Chen Y, Aldave AJ, Zhang F. Prevalence of transforming growth factor beta-induced gene corneal dystrophies in Chinese refractive surgery candidates. J Cataract Refract Surg. 2017;43:1489–94.

Souza JG, Dias K, Pereira TA, Bernardi DS, Lopez RF. Topical delivery of ocular therapeutics: carrier systems and physical methods. J Pharm Pharmacol. 2014;66:507–30.

Steinert SGAZMRF. Phototherapeutic Keratectomy. Am Acad Ophthalmol. 2013;12:2013.

Stenson PD, Mort M, Ball EV, Evans K, Hayden M, Heywood S, Hussain M, Phillips AD, Cooper DN. The Human Gene Mutation Database: towards a comprehensive repository of inherited mutation data for medical research, genetic diagnosis and next-generation sequencing studies. Hum Genet. 2017;136:665–77.

Sternberg SH, Redding S, Jinek M, Greene EC, Doudna JA. DNA interrogation by the CRISPR RNA-guided endonuclease Cas9. Nature. 2014;507:62–7.

Stone EM, Mathers WD, Rosenwasser GO, Holland EJ, Folberg R, Krachmer JH, Nichols BE, Gorevic PD, Taylor CM, Streb LM, et al. Three autosomal dominant corneal dystrophies map to chromosome 5q. Nat Genet. 1994;6:47–51.

Stewart HS, Ridgway AE, Dixon MJ, Bonshek R, Parveen R, Black G. Heterogeneity in granular corneal dystrophy: identification of three causative mutations in the TGFBI (BIGH3) gene-lessons for corneal amyloidogenesis. Hum Mutat. 1999;14(2):126–32. https://doi.org/10.1002/(SICI)1098-1004(1999)14:2<126::AIDHUMU4>3.0.CO;2-W. PMID: 10425035.

Stix B, Leber M, Bingemer P, Gross C, Rüschoff J, Fändrich M, Schorderet DF, Vorwerk CK, Zacharias M, Roessner A, Röcken C. Hereditary lattice corneal dystrophy is associated with corneal amyloid deposits enclosing C-terminal fragments of keratoepithelin. Invest Ophthalmol Vis Sci. 2005;46(4):1133–9. https://doi.org/10.1167/iovs.04-1319. PMID: 15790870.

Subrizi A, Del Amo EM, Korzhikov-Vlakh V, Tennikova T, Ruponen M, Urtti A. Design principles of ocular drug delivery systems: importance of drug payload, release rate, and material properties. Drug Discov Today. 2019;24(8):1446–57.

Taketani Y, Kitamoto K, Sakisaka T, Kimakura M, Toyono T, Yamagami S, Amano S, Kuroda M, Moore T, Usui T, Ouchi Y. Repair of the TGFBI gene in human corneal keratocytes derived from a granular corneal dystrophy patient via CRISPR/Cas9-induced homology-directed repair. Sci Rep. 2017;7:16713.

Takács L, Losonczy G, Matesz K, Balogh I, Sohajda Z, Tóth K, Fazakas F, Vereb G, Berta A. TGFBI (BIGH3) gene mutations in Hungary—report of the novel F547S mutation associated with polymorphic corneal amyloidosis. Mol Vis. 2007;18(13):1976–83. PMID: 17982422.

Tian X, Fujiki K, Wang W, Murakami A, Xie P, Kanai A, Liu Z. Novel mutation (V505D) of the TGFBI gene found in a Chinese family with lattice corneal dystrophy, type I. Jpn J Ophthalmol. 2005;49(2):84–8. https://doi.org/10.1007/s10384-004-0167-7. PMID: 15838722.

Tian X, Fujiki K, Zhang Y, Murakami A, Li Q, Kanai A, Wang W, Hao Y, Ma Z. A novel variant lattice corneal dystrophy caused by association of mutation (V625D) in TGFBI gene. Am J Ophthalmol. 2007;144(3):473–5. https://doi.org/10.1016/j.ajo.2007.04.015. PMID: 17765440.

Tsubota K, Inoue H, Ando K, Ono M, Yoshino K, Saito I. Adenovirus-mediated gene transfer to the ocular surface epithelium. Exp Eye Res. 1998;67:531–8.

Venkatraman A, Murugan E, Lin SJ, Peh GSL, Rajamani L, Mehta JS. Effect of osmolytes on in-vitro aggregation properties of peptides derived from TGFBIp. Sci Rep. 2020;10(1):4011. https://doi.org/10.1038/s41598-020-60944-0. PMID: 32132634; PMCID: PMC7055237.

Walton RT, Christie KA, Whittaker MN, Kleinstiver BP. Unconstrained genome targeting with near-PAMless engineered CRISPR-Cas9 variants. Science. 2020;368(6488):290–6. https://doi.org/10.1126/science.aba8853. Epub 2020 Mar 26. PMID: 32217751; PMCID: PMC7297043.

Warren JF, Abbott RL, Yoon MK, Crawford JB, Spencer WH, Margolis TP. A new mutation (Leu569Arg) within exon 13 of the TGFBI (BIGH3) gene causes lattice corneal dystrophy type I. Am J Ophthalmol. 2003;136(5):872–8. https://doi.org/10.1016/s0002-9394(03)00541-5. PMID: 14597039.

Weiss JS, Møller HU, Aldave AJ, Seitz B, Bredrup C, Kivelä T, Munier FL, Rapuano CJ, Nischal KK, Kim EK, Sutphin J, Busin M, Labbé A, Kenyon KR, Kinoshita S, Lisch W. IC3D classification of corneal dystrophies—edition 2. Cornea. 2015;34:117–59.

Wittebol-Post D, van der Want JJL, van Bijsterveld OP. Granular dystrophy of the cornea (Groenouw's Type I). Ophthalmologica. 1987;195(4):169–77. https://doi.org/10.1159/000309808.

Welch WJ, Brown CR. Influence of molecular and chemical chaperones on protein folding. Cell Stress Chaperones. 1996;1(2):109–15. https://doi.org/10.1379/1466-1268(1996)001<0109:iomacc>2.3.co;2. Erratum in: Cell Stress Chaperones 1996 Sep;1(3):207. PMID: 9222596; PMCID: PMC248462.

Wu X, Scott DA, Kriz AJ, Chiu AC, Hsu PD, Dadon DB, Cheng AW, Trevino AE, Konermann S, Chen S, Jaenisch R, Zhang F, Sharp PA. Genome-wide binding of the CRISPR endonuclease Cas9 in mammalian cells. Nat Biotechnol. 2014;32:670–6.

Yamamoto S, Okada M, Tsujikawa M, Shimomura Y, Nishida K, Inoue Y, Watanabe H, Maeda N, Kurahashi H, Kinoshita S, Nakamura Y, Tano Y. A kerato-epithelin (betaig-h3) mutation in lattice corneal dystrophy type IIIA. Am J Hum Genet. 1998;62(3):719–22. https://doi.org/10.1086/301765. PMID: 9497262; PMCID: PMC1376959.

Yang J, Han X, Huang D, Yu L, Zhu Y, Tong Y, Zhu B, Li C, Weng M, Ma X. Analysis of TGFBI gene mutations in Chinese patients with corneal dystrophies and review of the literature. Mol Vis. 2010;30(16):1186–93. PMID: 20664689; PMCID: PMC2901189.

Yavuz B, Kompella UB. Ocular drug delivery. Handb Exp Pharmacol. 2017;242:57–93.

Yellore VS, Rayner SA, Aldave AJ. TGFB1-induced extracellular expression of TGFBIp and inhibition of TGFBIp expression by RNA interference in a human corneal epithelial cell line. Invest Ophthalmol Vis Sci. 2011;52:757–63.

Yu P, Gu Y, Yang Y, Yan X, Chen L, Ge Z, Qi M, Si J, Guo L. A clinical and molecular-genetic analysis of Chinese patients with lattice corneal dystrophy and novel Thr538Pro mutation in the TGFBI (BIGH3) gene. J Genet. 2006;85(1):73–6. https://doi.org/10.1007/BF02728974. PMID: 16809844.

Yuan C, Zins EJ, Clark AF, Huang AJ. Suppression of keratoepithelin and myocilin by small interfering RNAs (siRNA) in vitro. Mol Vis. 2007;13:2083–95.

Zenteno JC, Ramirez-Miranda A, Santacruz-Valdes C, Suarez-Sanchez R. Expanding the mutational spectrum in TGFBI-linked corneal dystrophies: Identification of a novel and unusual mutation (Val113Ile) in a family with granular dystrophy. Mol Vis. 2006;10(12):331–5. PMID: 16636649.

Zenteno JC, Correa-Gomez V, Santacruz-Valdez C, Suarez-Sanchez R, Villanueva-Mendoza C. Clinical and genetic features of TGFBI-linked corneal dystrophies in Mexican population: description of novel mutations and novel genotype-phenotype correlations. Exp Eye Res. 2009;89(2):172–7. https://doi.org/10.1016/j.exer.2009.03.004. Epub 2009 Mar 18. PMID: 19303004.

Zhang C, Zeng G, Lin H, Li D, Zhao L, Zhou N, Qi Y. A novel mutation I522N within the TGFBI gene caused lattice corneal dystrophy I. Mol Vis. 2009;28(15):2498–502. PMID: 19956413; PMCID: PMC2786890.

Zhang Y, Sukthankar P, Tomich JM, Conrad GW. Effect of the synthetic NC-1059 peptide on diffusion of riboflavin across an intact corneal epithelium. Invest Ophthalmol Vis Sci. 2012;53(6):2620–9. https://doi.org/10.1167/iovs.12-9537. PMID: 22447859; PMCID: PMC3358128.

Zheng Q, Cai X, Tan MH, Schaffert S, Arnold CP, Gong X, Chen CZ, Huang S. Precise gene deletion and replacement using the CRISPR/Cas9 system in human cells. Biotechniques. 2014;57:115–24.

Zheng GX, Lau BT, Schnall-Levin M, Jarosz M, Bell JM, Hindson CM, Kyriazopoulou-Panagiotopoulou S, Masquelier DA, Merrill L, Terry JM, Mudivarti

PA, Wyatt PW, Bharadwaj R, Makarewicz AJ, Li Y, Belgrader P, Price AD, Lowe AJ, Marks P, Vurens GM, Hardenbol P, Montesclaros L, Luo M, Greenfield L, Wong A, Birch DE, Short SW, Bjornson KP, Patel P, Hopmans ES, Wood C, Kaur S, Lockwood GK, Stafford D, Delaney JP, Wu I, Ordonez HS, Grimes SM, Greer S, Lee JY, Belhocine K, Giorda KM, Heaton WH, McDermott GP, Bent ZW, Meschi F, Kondov NO, Wilson R, Bernate JA, Gauby S, Kindwall A, Bermejo C, Fehr AN, Chan A, Saxonov S, Ness KD, Hindson BJ, Ji HP. Haplotyping germline and cancer genomes with high-throughput linked-read sequencing. Nat Biotechnol. 2016;34:303–11.

Zinn E, Vandenberghe LH. Adeno-associated virus: fit to serve. Curr Opin Virol. 2014;8:90–7.

Zhong X, Chen S, Huang W, Yang J, Chen X, Zhou Y, Zhou Q, Wang Y. Novel and known mutations of TGFBI, their genotype-phenotype correlation and structural modeling in 3 Chinese families with lattice corneal dystrophy. Mol Vis. 2010;15(16):224–30. PMID: 20161820; PMCID: PMC2822555.

MIX
Papier aus verantwortungsvollen Quellen
Paper from responsible sources
FSC® C105338

If you have any concerns about our products,
you can contact us on
ProductSafety@springernature.com

In case Publisher is established outside the EU,
the EU authorized representative is:
Springer Nature Customer Service Center GmbH
Europaplatz 3, 69115 Heidelberg, Germany

Printed by Libri Plureos GmbH
in Hamburg, Germany